MINISTRY OF
AGRICULTURE, FISHERIES AND FOOD

Beneficial Insects and Mites

B. D. MORETON
B.Sc. Agric. (Lond.), Dip. Agric. (Wye), Dip. Ent. (Wye)
Advisory Entomologist, N.A.A.S. South-eastern Region

Bulletin 20

LONDON
HER MAJESTY'S STATIONERY OFFICE
1969

First published 1922
Sixth edition 1969

Bulletin 20: Beneficial Insects and Mites

CORRECTIONS

Page 21, para. 4, line 4. For "*Calosama*" read "*Calosoma*"

Page 24, para. 4, line 8. For "*succiola*" read "*succicola*"

Page 60, para. 3, line 3 should read "as demeton-methyl, but the more pesticidal component, demeton-S-methyl,"

Page 69, reference 12, line 2. For "*1965*" read "*1964*"

Page 69, reference 13, line 2. For "*1964*" read "*1965*"

Page 73, para. 1, line 5. For "was moth" read "wax moth"

Page 74, *Insecticides*, *Group II*, col. 3. For "Odydemeton-methyl" read "Oxydemeton-methyl"

Page 108, para. 4, line 7. For "flowers of fruit" read "flowers or fruit"

Ministry of Agriculture, Fisheries and Food
June 1970.

LONDON: HER MAJESTY'S STATIONERY OFFICE

Foreword

No one will deny that the continued development of pesticides over the past twenty-five years has strengthened the armoury at the disposal of the grower for the control of what are commonly called insects of economic importance. Equally it is undeniable that a variety of side effects of the use of pesticides has been detected, ranging from the development of resistance in some species, to the destruction of parasites and predators leading to the resurgence of hitherto minor pests. These facts, coupled with the increasing importance of minimizing pesticide residues and the need to keep down costs of production, have greatly emphasized the role of those other insects of economic importance—the beneficial insects—in pest control.

B. D. Moreton of the National Agricultural Advisory Service at Wye, who wrote the fifth edition of this Bulletin in 1958, has again undertaken its revision. This has been a major task involving the complete rewriting of many sections as a result of the vast amount of new work on the effects of pesticides on beneficial insects, the role of native parasites and predators and the renewed interest in the possibilities of biological control through the use of introduced species. In this connection the use of a South American predatory mite for the control of red spider mites in glasshouses is a stimulating example. Nor have the possibilities of integrated control—the rational use of pesticides coupled with biological control—escaped attention and the overall effect of this treatment has resulted in an edition that is more than double the size of the last with many references and suggestions for further reading. As before there are descriptive accounts of various beneficial insects, some of which are illustrated by water-colour drawings mainly done by Mr. Moreton, some of which are new.

Dr. J. B. Free and Dr. J. H. Stevenson of Rothamsted Experimental Station respectively contributed the material on pollinating insects and the effect of pesticides on them but the main credit for this edition goes to Mr. Moreton.

There is no doubt it will remain for many years the standard authority for growers and advisers alike, not forgetting all those who are concerned for the preservation of the diversity of our insect fauna in which beneficial insects are an important if not always a visually exciting element.

M. Cohen

Ministry of Agriculture, Fisheries and Food
Plant Pathology Laboratory
Harpenden

December 1968

Acknowledgments

THE author is greatly indebted to Dr. J. B. Free of Rothamsted Experimental Station for contributing the section on pollinating insects and to Dr. J. H. Stevenson, also of Rothamsted Experimental Station, for the section on the effects of pesticides on bees. Especial thanks are also gladly given to Miss Marion Gratwick of the Plant Pathology Laboratory, Harpenden, for her help and for the meticulous care she has taken in editing this Bulletin.

Acknowledgment of Illustrations

Drawings for Plates I–V

Twenty-three of the coloured drawings were supplied by Mr. B. D. Moreton and four by Mr. H. B. Foster. The remaining fifteen were redrawn by Mr. B. D. Moreton from the originals in the fourth edition.

Photographs for Plates VI–VIII

The eighteen colour photographs of insects were supplied by Dr. J. B. Free and the colour photograph of mites by Mr. J. R. Morrison.

Individual acknowledgments are given with the captions.

Editor's Note

ON the first occasion that an insect or mite species is mentioned in the text of this Bulletin, the Latin name and the authority are given together with the common English name if one exists; on subsequent occasions, the common name is used or, if there is no common name, the Latin name is used without the authority. The Latin names of all species are included in the index but readers are referred to the corresponding common names when these are used in the text.

Contents

Introduction

HONEY and silk are probably the only generally recognized benefits derived from insects. The silkworm was domesticated by neolithic man in China as long ago as the third millenium B.C., and familiarity with the produce of honey bees must be even older. Most other insects, apart perhaps from ladybird beetles and some of the attractively coloured butterflies, are commonly regarded with suspicion or fear, to be treated all alike as pests and destroyed without hesitation. This is to be regretted as only a few hundred of the 21,000 and more different kinds of insect in Great Britain could be listed as pests because they damage crops, livestock, stored food, or are carriers of disease, or attack man himself. A great many of the remainder live in the rich variety of environmental niches occurring over the face of the Earth without interfering with man's affairs, while there are also a considerable number whose activities in several indirect ways are of great benefit to him.

In passing it may be mentioned that, as well as the silkworm and honey bee, there are a few more insects of direct value. A number of substances derived from members of the scale insect and mealybug family have been used for a long time, such as kermes, the dyestuff prepared from dried females of *Kermes ilicis* (L.); shellac from *Laccifer lacca* (Kerr) in India, the females of which also yield the crimson lac dye; and wax from the Chinese *Ericerus pela* (Chav.). Cochineal dye is made from the Mexican *Dactylopius coccus* Costa, and a similar dye from a related species was known in biblical times. Some insects are eaten as food: for example, certain American Indians cook and eat large white beetle grubs that tunnel in trees. According to Leviticus, God instructed Moses and Aaron that locusts and grasshoppers, but no other insects, were permissible as food, and these are still accepted as a palatable item of food by natives of some countries. The wings of some brightly coloured butterflies are used for decorative purposes. Blowfly larvae ('gentles') are bred as bait for anglers, mealworms as food for caged birds, and ants' cocoons ('ants' eggs') are used to feed fish in aquaria. A remarkable use of blowfly larvae was the development of a treatment of the acute bone disease known as osteomyelitis, in which the larvae, reared under strictly sterile conditions, were placed in the lesions, where they cleared out the purulent material and dead tissues, enabling rapid healing to occur.

The main purpose of this Bulletin is to give some account of certain insects and mites whose activities, largely unnoticed by the casual observer, are of vital importance. These are pollinating insects, without which many flowers including crops could not set seed or produce fruit, and insects and mites which live at the expense of other insects and related creatures, either by hunting them down as food or living upon or inside them as parasites. Mention will also be made of some plant-feeding insects which have proved beneficial as weed control agents.

Predators and Parasites

ALL animal life depends ultimately upon plant life for its existence. Many animals, including insects, feed upon plants, while many more feed in some way or another on other animals. In any locality various forms of plant and animal life, which during the course of evolution have become adapted to the climatic and other prevailing conditions, live in complex inter-related associations. In these communities competition for the available food is keen; then, as Darwin (1859) put it, 'a struggle for existence inevitably follows from the high rate at which all organic beings tend to increase . . . there is no exception to the rule that every organic being naturally increases at so high a rate that, if not destroyed, the Earth would soon be covered by the progeny of a single pair . . . battle within battle must ever be recurring with varying success; and yet in the long run the forces are so nicely balanced that the face of nature remains uniform for long periods of time. . . .'

Several calculations have been made to illustrate the high reproductive capacity of some insects, especially with aphids (greenfly). During the summer, female aphids give birth to living young which are all females and these quickly mature to reproduce in the same way, thus giving rise to a succession of generations. The females of the peach-potato aphid (*Myzus persicae* (Sulz.)) can each produce 40 or more young and, if favourable conditions were maintained, one female could in 100 days be the ancestor of progeny weighing nearly ten million tons (Watson, 1967). With the house fly (*Musca domestica* L.), it has been estimated that the progeny of one pair, if all survived, could in one summer amount to 191 trillion.

Obviously such unchecked increase would ultimately be disastrous for all concerned. Outbreaks involving huge numbers of insects do occur from time to time, even in the temperate British Isles. In occasional seasons caterpillars of the green oak tortrix moth (*Tortrix viridana* (L.)) are so abundant in some localities as to defoliate oak trees; periodically leather-jackets, the larval or grub stage of the familiar daddy longlegs, are sufficiently numerous to cause severe damage to cereal crops and even grassland, while at other times they are almost forgotten. Caterpillars of the antler moth (*Cerapteryx graminis* (L.)), which feed on grasses, sometimes occur in vast numbers, especially in the northern parts of the country, denuding areas of grassland; after heavy rain in hilly districts, drains and ditches may become filled with countless thousands of these caterpillars. Migrant insects sometimes give rise to plagues, as did the diamond-back moth (*Plutella xylostella* (L.)) in 1958, when a massive invasion of moths, which had apparently come from Finland or north-west Russia, occurred in north-east England and eastern Scotland in June, and this was followed by destructive infestations of caterpillars on brassica crops. In hotter climates the appearance of vast numbers of insects, with disastrous effects on crops, are well known and even more dramatic.

Variations from year to year in insect numbers, whether of pests or harmless kinds, are a matter of common observation but, in spite of their great powers of reproduction, insect populations are generally held to a fluctuating though comparatively low level. Every organism has certain specific needs, and optimum environmental conditions for it do not exist everywhere or at all times. Where they do exist the organism can multiply to its capacity, but sooner or later its progress will be arrested or reversed, owing to such factors as depletion of its food supply and accumulation of waste products, to adverse weather, to diseases, and to the effects of predatory and parasitic organisms. Insect populations, like those of other animals, are all subject to regulating forces of this kind. One or other may be the dominant factor at certain times, or certain stages in the insect's life cycle, and as some of the factors are often influencing the effectiveness of others, a complex natural control mechanism is brought about.

The predatory organisms may be many and various, including mammals, birds, reptiles, amphibians, spiders, centipedes, mites, and other insects. Insects feeding on other live animals may be classified into a number of types according to their habits: blood-suckers or those that take an occasional meal of blood, e.g., mosquitoes and horse flies; predators, which are those that hunt out and devour their prey, attacking a number of individuals during their lives, e.g., ladybird grubs and ground beetles; and parasites or those that live on and at the expense of a single individual called the 'host'.

Parasites may be sub-divided into two groups, i.e., those which live on or in the bodies of vertebrate hosts without causing their death, and fatal parasites (parasitoids), whose hosts are relatively small and are usually insects. Examples of the first kind are lice, fleas and bed bugs, which live on the bodies of mammals and birds, where they eat skin, hair or feathers, or suck blood. The curious-looking insects called *Stylops* can, usually with the exception of the males, spend all their lives within the bodies of bees, wasps or certain leafhoppers without killing their host. Most parasitic insects of the second group, however, are parasitic only in their young or larval stages when they feed on or inside the body of their host, eventually bringing about its death; they leave the devouring of their host's vital organs until they have practically completed their own development.

The habit of parasitism is frequently highly specialized for, although some parasites attack several different kinds of host, many have precise requirements restricted to particular species. Some attack other parasitic insects and are known as hyperparasites. The habits of the young stages, living in the fluid medium of the host's body in such a way as to prevent its premature illness or death, obviously entail special adaptation; most of the parasitic insects are found in the two large orders, Diptera (flies) and Hymenoptera (bees, wasps and ants). Predatory insects, on the other hand, require little modification of structure compared with plant-eating kinds, and examples of predators are found throughout the insect orders.

Probably few, if any, of the many species of insect escape attack by either predatory or parasitic insects or both. While the effects of these must vary considerably according to circumstances and are not the only natural means of regulating the size of insect populations, they are often of the greatest importance. Compared with physical factors, such as adverse weather, which probably destroy the same proportion of an insect population whatever its size, predators and parasites tend to become more effective as the host population increases. As more host insects become readily available their predators and parasites are likely to multiply; then, as through their activities the hosts become scarcer, the numbers of predators and parasites may decline through starvation or dispersal to fresh areas. Surviving host insects can then breed again more freely until once more overtaken by their enemies. Sequences of this sort explain some of the fluctuations in insect numbers, though of course they are likely to be subject to modification by the subtle interplay of other factors.

The potential importance of predators and parasites is well illustrated by events following the introduction of some insect and mite pests into new countries. Before the dangers were realised, many insects were accidentally taken from Europe and Asia to North America, where they became far worse pests than they were in their homeland. More favourable climatic and other conditions were doubtless sometimes or in part responsible, but the successful control following deliberate introduction of some of their parasites or predators suggests that absence of their native natural enemies was often the chief reason. Examples of control of introduced pests by such means are described on pages 90–94.

Insects have not neglected to exploit the immense variety of modes of life which can be followed on this Earth, and a great many kinds have become predators or parasites. The better known ones found in this country are described in the following pages. Much more knowledge of them is needed, and problems produced by the use of insecticides in recent years (referred to on pages 51–66) have stimulated further research.

References and Further Reading

DARWIN, CHARLES (1859). *On the origin of species by means of natural selection, or the preservation of favoured races in the struggle for life.* London, John Murray.

DEBACH, PAUL (Ed.) (1964). *Biological control of insect pests and weeds.* London, Chapman and Hall.

ORDISH, GEORGE (1967). *Biological methods in crop pest control.* London, Constable.

SWAN, LESTER A. (1964). *Beneficial insects.* New York, Harper and Row.

WATSON, MARION A. (1967). Epidemiology of aphid-transmitted plant-virus diseases. *Outl. Agric.* **5**, 155–66.

CAPSID AND RELATED BUGS

The order Hemiptera comprises a large and important group of insects with over 1,400 species in Britain. They are characterized by the formation of the mouthparts into a slender and jointed beak (proboscis), which when not in use lies along the underside of the head and thorax. There are usually two pairs of wings; the fore pair is generally harder than the membraneous hind pair, either uniformly so (sub-order Homoptera) or, as in the predatory representatives to be considered below, with the basal portion harder and thicker than the remainder (sub-order Heteroptera). These modified wings are not, however, veinless and horny as in the beetles (page 14).

The Hemiptera do not have a pupal (chrysalid) stage in their growth from egg to adult, and the young on hatching from the egg resemble the adult in general appearance except that they are smaller and lack wings. Young of this kind are commonly called nymphs; with each change of skin as they increase in size they become more like the adult, the wings appearing as 'buds' on the thorax after the third moult.

The majority of the Hemiptera are plant feeders, using their mouthparts to puncture plant tissues and extract food material. Many, such as the aphids and scale insects, are important pests of crops; others, for example bed bugs, feed on animal blood, while some, such as the black-kneed capsid, are predatory on other insects and on mites.

Capsid Bugs

These are members of the family Miridae, formerly known as Capsidae from which was derived the common name of 'capsid bugs'. Some 185 species are known in Britain; many of these are plant feeders and include pests such as the apple capsid (*Plesiocoris rugicollis* (Fall.)) and common green capsid (*Lygocoris pabulinus* (L.)), but several are important predators. Much of our knowledge of these predators is due to a study of the natural enemies of fruit tree red spider mite (*Panonychus ulmi* (Koch)) begun at East Malling Research Station in 1945 (Collyer 1952, 1953a, 1953b, 1953c). In this study, which was made in both neglected, unsprayed orchards and commercial, regularly sprayed orchards in Kent and Essex, 16 species of capsid bug predatory on fruit tree red spider mite were recorded.

Black-kneed Capsid

The most important capsid bug, since it was the most numerous in the cultivated orchards studied, was the black-kneed capsid (*Blepharidopterus angulatus* (Fall.)) (Collyer, 1952). Adults of this species are about $\frac{3}{16}$ in. (5 mm) in length; the males are rich green with dark red eyes while the females are yellowish-green with paler eyes. In both sexes there is a narrow, black band at the base of the long segment (tibia) of each leg (Plate I); these black 'knees' distinguish this insect from nearly all other capsids (see page 5). The nymphs are yellow to green with red eyes and black 'knees'.

The bugs lay their eggs from July to October, but chiefly in August, on a wide variety of trees. On apple the eggs are laid principally on one- or two-year-old shoots The female prefers the softer tissues, sometimes making use of bud-scale scars, lenticels (breathing pores in the bark) and small wounds, but more frequently she makes special slits in which to bury her eggs. Favourite sites are at the base and around the lower buds of the young shoots. The egg is whitish, long and curved, with a cap at one end from which the nymph eventually emerges, and is usually sunk in the bark so that only this cap is visible (Plate I). When eggs are laid in smooth wood the injury to the shoot results, after two or three days, in the development of a characteristic small swelling which has been described as an 'egg bump'.

The eggs are usually deposited singly and, under laboratory conditions, up to 50 are laid by each female. They hatch in the following year at times varying with the season but generally from towards the end of May to July or even August, with a peak near the end of June. At hatching, the cap of the egg is pushed to one side and, within five minutes of its escape, the nymph is running actively in search of food. The nymphal period varies from 25–35 days, adults beginning to appear in July. The females may remain active on the trees until October as long as leaves and mites are present. There is only one generation in the year.

When searching for food, the nymphs move unsystematically but swiftly backwards and forwards over the leaves. Both nymphs and adults are equally predatory and have a preference for the fruit tree red spider mite, particularly the adult female. The victims are sucked dry and the empty skins left on the leaf; the summer eggs of the mite are occasionally emptied but rarely the winter eggs. The bugs sometimes feed on leaves and fruitlets but cannot survive on a vegetarian diet alone. In a laboratory investigation the average total of adult female fruit tree red spider mites consumed throughout the season by each female bug in 1947 was 3,248, and in 1948 was 4,230. The nymphs accounted for some 5–40 a day according to their age, and the adult females an average of 60–70 a day. Both adults and nymphs will also attack various other mites, small insects such as leafhoppers, aphids, thrips and caterpillars, and other predators including their own species.

The black-kneed capsid is evidently capable of making a marked impression on infestations of fruit tree red spider mite, and further work (Muir, 1965) has shown that, provided no insecticides injurious to the predator are used, it can stabilize fruit tree red spider mite populations. The interaction between the predator and its prey, and the biological control of the pest which is achieved, are described on page 83 *et seq*.

Other Capsid Bugs

The other 15 species studied in this investigation (Collyer, 1953a, 1953b, 1955) can only be briefly mentioned. *Campylomma verbasci* (Mey.–Dür), unlike the others, has two generations a year, the winter eggs hatching in May and the summer eggs from the end of July to the end of September; adults may be present until late October. This bug was found abundantly in cultivated orchards in Essex and Kent but was rare in the neglected orchard studied. It feeds chiefly on fruit tree red spider mite but occasionally on aphids. *Camptobrochis lutescens* (Schill.) hibernates as an adult, emerging April–May and laying eggs for an extended period in the summer; it feeds on aphids and mites until October. It occurred regularly in small numbers in both the neglected and cultivated orchards and was found on many trees and shrubs as well as on fruit.

All the other species overwinter as eggs which are buried in the young wood, except *Malacocoris chlorizans* (Panz.) which places them on the outside of the bark. Those remaining to be noted were usually found more often in the neglected orchard than in cultivated ones, and all occurred on a variety of plants in addition to fruit, though the black apple capsid (*Atractotomus mali* (Mey.–Dür)) was more frequent on apple. *Psallus ambiguus* (Fall.), the red apple capsid (Plate I), occurred plentifully and was the earliest to appear, hatching from early April to early May, the adults completing their egg laying and disappearing by the end of June. The black apple capsid was abundant, hatching in May and occurring until July. *Plagiognathus arbustorum* (F.), pale yellow to almost black, likewise hatches in May at the same time as the black-kneed capsid; it also has black knees but with additional black spots on the same leg segment, black-margined thighs (femora) and black bands on the antennae. It was occasionally plentiful and adults were seen from July to September. *Campyloneura virgula* (Herr.–Schäff.) hatches over a long period from early May to August. *Deraeocoris ruber* (L.) hatches from mid-May to late June and is found up to the end of August.

Others appearing in May were *Orthotylus marginalis* Reut., the dark green apple capsid, and *O. nassatus* (F.), both persisting into August and sometimes plentifully. Several other species with later hatching periods also occurred. *Heterotoma planicornis* (Pall.) (=*Capsus meriopterus* (Scop.)), a species with the first two antennal segments conspicuously thickened, hatches from late May to September; it was found in small numbers and less commonly in neglected than in commercial orchards. *Pilophorus perplexus* Doug. & Scott and *M. chlorizans* hatch in June and adults occur until September or October; the former occurred in small numbers while the latter was abundant. *Phytocoris tiliae* (F.), *P. ulmi* (L.) and *P. reuteri* Saund. hatch from June to August and adults are present until October. *P. tiliae* was found in both commercial and neglected orchards but more abundantly in the latter, while *P. ulmi* and *P. reuteri* were less common, apple being only an occasional host plant.

All these capsid bugs feed on mites, including fruit tree red spider mite, and many of them also feed on aphids and other small insects. *P. tiliae, H. planicornis* and the black apple capsid include caterpillars in their prey, and *D. ruber*, which prefers larger prey, is cannibalistic and attacks other predacious species. Several of these capsid bugs also occasionally feed on plants; on apple they puncture leaves and fruit.

ANTHOCORID BUGS

Another family, the Anthocoridae, includes some important predators, and these were also studied in the investigation mentioned above (Collyer, 1953a). *Anthocoris nemorum* (L.) (Plate I) occurred regularly and abundantly on fruit trees in neglected and cultivated orchards. The adults, usually only the females, hibernate under bark, leaves, in hedgerows, etc., emerging in March and April and often congregating on willow catkins. They are coloured black and brown and are about $\frac{1}{6}$ in. (4 mm) long. Eggs are laid from the end of March until June and a new generation of adults begins to appear in May, all stages of the insect being present until September. The eggs are laid beneath the epidermis of leaves, usually on the undersurface, with the white circular cap protruding; one female may lay up to 200 eggs. The nymphs are a dark reddish-brown (Plate I).

The prey of this bug includes aphids, scale insects, apple sucker (*Psylla mali* Schmid.), capsid bugs, caterpillars, midges, apple blossom weevil (*Anthonomus pomorum* (L.)) larvae and pupae, and mites. The winter eggs of fruit tree red spider mite are attacked in autumn and spring and on warm days in winter. The adults were found to suck an average of about 50 fruit tree red spider mites a day, while in other investigations (Peska, 1931) some 500–600 aphids were consumed by one individual in its lifetime. Two related species, *A. nemoralis* (F.) and *A. confusus* Reut., which have similar habits, also occurred occasionally in both types of orchard.

Two more members of this family were found to be plentiful both in neglected and cultivated orchards (Collyer, 1953a): *Orius majusculus* (Reut.), which was the more numerous, and *O. minutus* (L.) (Plate I). *O. majusculus* is only half the length of *A. nemorum* and the young stages are light brown. They also overwinter as adults, mainly as females, which emerge in March and April. There are two generations a year, all stages occurring from April to August, and eggs are laid on the midribs on the underside of the leaves. The adults were found to eat about 35 adult fruit tree red spider mites a day and, like *A. nemorum*, also sucked the winter egg of the mite.

Anthocorid bugs are important predators of the lettuce root aphid (*Pemphigus bursarius* (L.)) (Dunn, 1960). This aphid feeds on the roots of outdoor lettuce and may cause considerable harm, especially in hot, dry weather. It overwinters as eggs on poplar, where in spring the young live in swollen, reddish galls on the leaf stalks; these galls split open in July to release winged aphids which disperse to lettuce. Anthocorid bugs will enter these open galls, usually one nymph per gall, and the entire aphid contents (which may average 100 individuals and may exceed 200) are rapidly killed before the bug starts feeding. It has been suggested that this remarkable effect may be due to fumigation by a secretion from the stink glands with which the bugs are furnished.

At Wellesbourne, Warwickshire, between 90 and 100 per cent of the galls on several poplar sites were found to have been attacked after mid-July in 1956, i.e., about 14 days after the beginning of the opening of the galls. Galls as high up as 16 ft, or even 35–40 ft on some trees, were affected to the same extent. It seems likely that the adult bugs lay their eggs outside the gall, the young subsequently penetrating the opening.

Two species, *A. nemorum* and *A. nemoralis*, were found in these galls, and *Orius minutus* has also been recorded. An interesting cycle of abundance was found at Wellesbourne over a period of four years. In 1955 the galls were very abundant while the anthocorid bugs were scarce initially, but on the plentiful food supply the anthocorids were able to build up their numbers so that a much increased population overwintered on the poplars. In 1956 galls were fewer, but still common; the anthocorids increased still further and their intense attack severely curtailed dispersal of the aphids. In 1957 the predators were numerous and galls were few; these were quickly entered and after the first week of July no galls containing live aphids could be found, so that the duration and strength of the aphid dispersal was considerably reduced. In the following year galls were extremely few, and were not attacked, possibly because in the previous year the predators found insufficient food and died of starvation or dispersed elsewhere.

In experimental plots of oats at Rothamsted Experimental Station, *A. nemorum* was common in 1966, occurring between the glumes in the oat panicles where it fed on all stages of the frit fly (*Oscinella frit* (L.)) (Jones, 1967).

Common black ants (*Lasius niger* (L.)) will protect colonies of black bean aphid (*Aphis fabae* Scop.) from predators, driving off both adult and immature *Anthocoris nemorum* as well as ladybird larvae (Banks, 1962).

Nabid Bugs

Himacerus (=*Nabis*) *apterus* (F.) and *Aptus mirmicoides* (Costa) (=*Himacerus lativentris* (Boh.)) (family Nabidae) may also be mentioned as predatory bugs feeding on fruit tree red spider mite, although they prefer larger prey (Collyer, 1953b). The former passes the winter as eggs laid deep in young apple wood, hatching occurs in May and June and adults are present until October, while the latter species overwinters as an adult and lays eggs early in the summer. Both occur on a variety of trees, being occasionally found on fruit, more especially in neglected orchards.

Other Predatory Bugs

There are predators from further families of bugs such as the Pentatomidae (shield bugs) and Reduviidae. The forest bug, *Pentatoma rufipes* (L.), which is sometimes numerous, especially in old cherry orchards, hibernates as a nymph and in the spring preys upon caterpillars of the winter moth group. The sloe bug, *Dolycoris baccarum* (L.), with a preference for plum and damson orchards, attacks aphids (Massee, 1954). The reduviid bugs are usually large and active hunters preying upon small insects, including gnats and midges, harvest spiders, etc.

References and Further Reading

Anderson, N. H. (1962). Bionomics of six species of *Anthocoris* (Heteroptera: Anthocoridae) in England. *Trans. R. ent. Soc. Lond.* **114,** 67–95.

Banks, C. J (1962). Effects of the ant *Lasius niger* (L.) on insects preying on small populations of *Aphis fabae* Scop. on bean plants. *Ann. appl. Biol.* **50,** 669–79.

Collyer, Elsie (1952). Biology of some predatory insects and mites associated with the fruit tree red spider mite (*Metatetranychus ulmi* (Koch)) in south-eastern England. I. The biology of *Blepharidopterus angulatus* (Fall.) (Hemiptera-Heteroptera, Miridae). *J. hort. Sci.* **27,** 117–29.

Collyer, Elsie (1953a). Biology of some predatory insects and mites associated with the fruit tree red spider mite (*Metatetranychus ulmi* (Koch)) in south-eastern England. II. Some important predators of the mite. *J. hort. Sci.* **28,** 85–97.

Collyer, Elsie (1953b). Biology of some predatory insects and mites associated with the fruit tree red spider mite (*Metatetranychus ulmi* (Koch)) in south-eastern England. III. Further predators of the mite. *J. hort. Sci.* **28,** 98–113.

Collyer, Elsie (1953c). Biology of some predatory insects and mites associated with the fruit tree red spider mite (*Metatetranychus ulmi* (Koch)) in south-eastern England. IV. The predator-mite relationship. *J. hort. Sci.* **28,** 246–59.

Collyer, Elsie (1955). Some predacious insects of fruit trees: capsid bugs. *Rep. E. Malling Res. Stn for 1954,* 155–59.

Collyer, Elsie (1967). On the ecology of *Anthocoris nemorum* (L.) (Hemiptera-Heteroptera). *Proc. R. ent. Soc. Lond.* (*A*) **42,** 107–18.

Dunn, J. A. (1960). The natural enemies of the lettuce root aphid, *Pemphigus bursarius* (L.). *Bull. ent. Res.* **51,** 271–78.

Jones, Margaret G. (1967). Predators of frit fly. In: Johnson, C. G. Entomology department. *Rep. Rothamsted exp. Stn for 1966,* 203–4.

Massee, A. M. (1954). *The pests of fruits and hops.* London, Crosby Lockwood.

Muir, R. C. (1965). The effect of sprays on the fauna of apple trees. I. The influence of winter wash, captan, and lime-sulphur on the interaction of populations of *Panonychus ulmi* (Koch) (Acarina:

Tetranychidae) and its predator, *Blepharidopterus angulatus* (Fall.) (Heteroptera:Miridae). *J. appl. Ecol.* **2,** 31–41.

PESKA, W. (1931). Obserwacje nad biologja dzióbalka gajowego (*Anthocoris nemorum* L.). *Pr. Wydz. Chorob Szkod. Rosl. panst. Inst. nauk. Gospod. wiejsk. Bydgoszczy*, No. 10, 53–71.

SOUTHWOOD, T. R. E. and LESTON, DENNIS (1959). *Land and water bugs of the British Isles.* The Wayside and Woodland Series. London and New York, Frederick Warne.

THRIPS

The Thysanoptera is an order with some 180 British representatives; they are small or minute, narrow-bodied insects, and wings, when present, are four in number, narrow, strap-shaped and fringed with long hairs.

The eggs of some thrips are laid in slits cut in plant tissues by the female's saw-like ovipositor, and those of others are laid on stems, leaves, or under bark. The newly hatched young resemble the adults in general form but lack wing-pads and have fewer antennal segments. In one of the two sub-divisions of the order a pupal and pre-pupal stage follow after the young have moulted twice, and in these the insect has wing-pads and does not feed. In the other sub-division an additional and inactive pupal stage occurs. Pupation usually takes place in the soil, though in some species it occurs on the host plant. The number of generations in the year differs according to the species, and hibernation may take place in the nymphal, pupal or adult stage.

Some are crop pests, for example the pea thrips (*Kakothrips robustus* (Uzel)), but a number are predacious insects. *Thrips tabaci* Lind., the onion thrips, generally regarded only as a plant feeder, has been found, together with *Aeolothrips melaleucus* Hal., also to attack the young stages of the fruit tree red spider mite (Collyer, 1953). *A. ericae* Bagn., *Haplothrips* species and *Scolothrips longicornis* Pries. are also predacious, the latter probably attacking spider mites.

REFERENCES AND FURTHER READING

COLLYER, ELSIE (1953). Biology of some predatory insects and mites associated with the fruit tree red spider mite (*Metatetranychus ulmi* (Koch)) in south-eastern England. III. Further predators of the mite. *J. hort. Sci.* **28,** 98–113.

MORISON, G. D. (1947–49). Thysanoptera of the London area (London thrips). *Lond. Nat.* 1946, No. 26, Suppl. 1–36; 1947, No. 27, Suppl. 37–75; 1948, No. 28, Suppl. 77–131. (Re-issued as 'London Naturalist' Reprint No. 59.)

DRAGONFLIES

These insects, belonging to the order Odonata, are too familiar to need introduction. They are in no way harmful to man and his crops, despite some popular beliefs, and they can be included among the beneficial insects because in their young stages passed in ponds and streams, and as adult dragonflies, they eat other insects. There are just over 40 British species.

The adults are sun-loving and are not confined to the proximity of water.

Many species are very swift in flight, though most do not travel far from their chosen haunts. Their large compound eyes indicate well-developed powers of vision, and they capture their prey while on the wing, grasping large victims with their long, forwardly directed legs. Their prey includes mosquitoes and other flies, beetles and various Hymenoptera.

In some species the eggs are dropped into the water, in others they are attached to plants, or laid in slits cut in plant stems near or below the surface of the water. The nymphs have elongate bodies, large compound eyes and three pairs of long legs. The lower mouthpart is modified to form a long tongue-like organ, armed at the tip with a pair of movable hooks and hinged at the base so that it can be folded back below the head. At rest this organ conceals the other mouthparts and is known as the 'mask'; it is used for catching prey and is shot forward with lightning rapidity to impale a victim on its hooks. The prey includes nymphs of mayflies, dragonflies (their own and other species) and various other aquatic insects; nymphs of the larger species will also attack tadpoles and even small fish. When fully grown the nymph crawls up a plant stem out of the water, its skin splits down the back and the adult emerges, leaving the empty skin hanging on its support.

Further Reading

Corbet, Philip S., Longfield, Cynthia, and Moore, N. W. (1960). *Dragonflies*. New Naturalist Series, No. 41. London, Collins.

Longfield, Cynthia (1949). *The dragonflies of the British Isles*. The Wayside and Woodland Series. London and New York, Frederick Warne.

LACEWINGS

Lacewings belong to the order Neuroptera which includes 53 species in Britain. They are small to fairly large insects with four transparent wings netted with a great many veins and cross-veins, of which a ladder-like series along the front margin of the fore wings is conspicuous. The hind and fore pairs are very similar and are laid roof-wise over the insect's back when it is resting. The antennae are long and thread-like.

Like many other groups of insects, such as butterflies and moths, the lacewings have egg, grub or larval, and chrysalid or pupal stages in their life cycle before becoming adult winged insects. The larvae, which have the general shape shown in Plate II, are active insects with six slender legs and a pair of slender, curved jaws like hollow callipers. With these jaws they seize aphids and various other small soft-bodied insects and suck their body fluids.

There are several families of lacewings, some of which have larvae living in water and damp situations; we are concerned with three families whose members live on land.

Green Lacewings

The green lacewings (family Chrysopidae), with 13 species, are fairly large with bright green or yellowish-green bodies and wings (excepting two species that are brown); they are also known as 'golden-eyes' owing to the metallic lustre of their compound eyes. They are slow fliers and are often attracted to lights. *Chrysopa carnea* Steph. (Plate II), the only British species known to hibernate as an adult, will enter houses to hibernate, when it changes from green to a reddish colour, resuming its green in the spring. Some species have a pair of glands on the front of the thorax from which they can emit a penetrating and unpleasant odour as a defensive measure if they are handled.

The greenish eggs are laid on stalks in batches (Plate II); the female first deposits a drop of gummy fluid on the leaf and, by rasing her abdomen, draws this up into a fine stalk which immediately hardens, and then places an egg on its summit. In *C. carnea* the egg stalks are separate, but in some species they are united near the base into a bundle. The larvae are bristly, usually whitish, green or yellowish, with red, brown or black markings. Some of them cover their back with the empty skins of the aphids and other insects they have attacked, so that they easily escape notice. When fully fed they spin a rounded silken cocoon (Plate II) in which they pass the pupal stage, and the adult lacewing emerges in due course. Some species hibernate as partially grown larvae, but the winter is generally passed as full-grown larvae in cocoons. There may be two broods in the year.

In laboratory experiments (Sundby, 1966) it was found that *C. carnea* laid an average of nearly 500 eggs at 21°C (70°F) over a period of about 17 days. The egg, larval and pupal stages occupied about 35 days. The percentage of eggs hatching was about 67, so that with two generations a year the progeny of one female could amount to nearly 53,000 during the summer. The average number of peach-potato aphids eaten by one larva in 15 days at 21°C (70°F) was nearly 400. Young larvae took about 16 minutes to eat one aphid, and fully grown larvae only 1½ minutes. It was estimated that the larvae produced during the summer from one female, if all survived, could consume over 20,000,000 aphids. The adult lacewings fed on honey and pollen, and did not lay eggs if given only an aphid diet. Their eggs appear to be laid scattered over plants without regard to the presence of aphid colonies; the newly hatched larvae will suck eggs of their own species.

Green lacewing larvae may be seen with aphid colonies on farm and garden crops, though they are usually not as frequent as larvae of ladybirds and hover flies. Larvae of *Chrysopa ventralis* Curt. (=*C. prasina* Burm.) are common on oak and beech, and in pine plantations where they exert a valuable check on adelges, which they continue to attack in the winter months. *C. flava* (Scop.) prefers large aphids, suckers and leafhoppers, even attacking beetles and spiders. *C. albolineata* Kill. and *C. septempunctata* Wesm. prey on the woolly aphid (*Eriosoma lanigerum* (Hsmnn.)) on apple trees; and the latter, with *C. phyllochroma* (Wesm.), also attacks the black bean aphid. *C. carnea* includes the fruit tree red spider mite as well as aphids in its prey (Collyer, 1953b).

Larvae of *C. carnea* are very swift and active predators. With the aid of their long jaws they seek out small aphids hidden in the folds of young leaves where ladybird larvae often fail to reach them. As with ladybirds, ants will also completely protect black bean aphids against lacewing larvae, which when threatened roll up and drop from the plants (Banks, 1962).

Brown Lacewings

The 29 species of brown lacewing (family Hemerobiidae) in Britain are generally smaller insects, with brown or greyish wings. They and their larvae, like the green lacewings, feed on aphids, thrips, mites, etc. The eggs are not laid on stalks, and the larvae, though of similar shape to those of the green lacewings, are not bristly and their jaws are shorter, stouter and less curved. They do not cover themselves with the skins of their victims or other debris. Some species occur in pine woods, where they thrive on aphids that are pests of coniferous trees. *Hemerobius humulinus* L. and *H. lutescens* F. prey on the fruit tree red spider mite (Collyer, 1953b). *Kimminsia* (=*Boriomyia*) *subnebulosa* (Steph.) is very common, its larva feeding on the black bean aphid (Withycombe, 1924); it was also noted among the predators of aphids on potatoes in northern England (Dunn, 1949).

Powdery Lacewings

The powdery lacewings (family Coniopterygidae) (Plate I), of which seven species are listed for Britain, are small fragile insects and unlike the other members of the Neuroptera in appearance. The wings are much less netted with veins, and wings and body are covered with a white powdery substance so that they somewhat resemble the glasshouse whitefly (*Trialeurodes vaporariorum* (Westw.)). In contrast to the whiteflies they hold their wings in a steep roof-wise manner when at rest. The larvae have plumper bodies than those of other lacewings, tapering to the tip of the abdomen, and long legs and short stout jaws. They prey upon small aphids, scale insects and mites.

A considerable amount was added to our knowledge of these insects by the work at East Malling Research Station on the predators of the fruit tree red spider mite (Collyer, 1951, 1953a). Adult lacewings of the species *Conwentzia pineticola* End. emerge from white silken cocoons on or beneath the bark of fruit trees in April and can be found in orchards until November, there being more than one generation in the year. The very small eggs are long and oval, salmon-pink in colour, and are laid singly on bark and on the undersides of leaves from April to June. After laying her egg the female covers it with a patch of white mealy substance similar to that clothing her body and wings. Further eggs are laid mainly in July and August by the second brood of adults. A partial third generation also occurs and this will lay eggs until as late as November. The lacewings are comparatively long-lived, the females living for up to eight weeks in the laboratory and probably longer out of doors. The highest number of eggs laid by one female in captivity was 107 in a period of seven weeks in September and October.

The larvae are more or less colourless when newly hatched, except for their red eyes, but when they feed, their body contents become deep red or brown so that they appear red with white markings (Plate I). A sticky substance is secreted by the last abdominal segment, forming a means of anchorage to a leaf; this is used especially by young larvae. When fully fed the larvae spin a conspicuous white silken and double-walled cocoon in which to transform to the pupal stage, the cocoon usually being spun on a leaf in summer but on or under bark in the autumn.

In orchards the larvae and adults feed almost entirely on mites and their eggs, although larvae have been known to feed on each other and on eggs of their own species. The larvae pierce fruit tree red spider mites with their jaws and suck out the body contents, leaving the empty skin on the leaf, and suck the mite's eggs, leaving an empty shell. They have a preference for adult female mites, which even newly hatched larvae, little bigger than the mites, will attack, and they have been observed to eat an average of 15–35 mites per day according to age. Larvae may be active in the autumn after the leaves have fallen, when they live almost entirely on the winter eggs of the mites; the colourless empty shells on the bark of fruit trees are evidence of their activity, which in some seasons effects a big reduction, as much as 90 per cent, in the numbers of live eggs. The adults pick up and completely devour mites, and when eating the contents of eggs bite off the top half of the shell leaving behind the lower half; they have been seen to eat 30 adult female mites in an hour.

C. pineticola was found in Essex on fruit trees of all kinds, but solely in association with fruit tree red spider mite infestations, and very occasionally on neighbouring hedges. It appears to occur spasmodically and either sparsely or plentifully; for example, in 1947 it was numerous in the late summer in many Essex orchards, but apparently the lacewings emerging in the following spring left for other localities; large numbers were seen again in Essex in the late summer of 1949. This species was also found in Kent.

A related lacewing, *C. psociformis* (Curt.), is known to occur on fruit and other trees and bushes, particularly hawthorn. It has two generations a year and is not active after September. It also preys on the fruit tree red spider mite.

Two other lacewings, *Coniopteryx tineiformis* Curt. and *Semiadalis aleyrodiformis* (Steph.) are found on fruit and other trees in small numbers. The larvae of the latter are distinctively black and white, and both have similar feeding habits to those already described.

The use of *Coniopteryx* species to control red spider mites on glasshouse crops was suggested by Withycombe (1924) but never tried. He also introduced cocoons or adults to various fruit trees and thereby succeeded in obtaining a marked reduction of mite infestations, but in the following year a parasite (a species of *Lygocerus*) appeared, to the detriment of the lacewing.

References and Further Reading

BANKS, C. J. (1962). Effects of the ant *Lasius niger* (L.) on insects preying on small populations of *Aphis fabae* Scop. on bean plants. *Ann. appl. Biol.* **50,** 669–79.

CLAUSEN, C. P. (1940). *Entomophagous insects.* New York and London, McGraw-Hill.

COLLYER, ELSIE (1951). The separation of *Conwentzia pineticola* End. from *Conwentzia psociformis* (Curt.), and notes on their biology. *Bull. ent. Res.* **42,** 555–64.

COLLYER, ELSIE (1953a). Biology of some predatory insects and mites associated with the fruit tree red spider mite (*Metatetranychus ulmi* (Koch)) in south-eastern England. II. Some important predators of the mite. *J. hort. Sci.* **28,** 85–97.

Collyer, Elsie (1953b). Biology of some predatory insects and mites associated with the fruit tree red spider mite (*Metatetranychus ulmi* (Koch)) in south-eastern England. III. Further predators of the mite. *J. hort. Sci.* **28,** 98–113.

Collyer, Elsie (1956). Some predacious insects of fruit trees: lacewings, ladybirds and hover flies. *Rep. E. Malling Res. Stn* for 1955, 155–56.

Dunn, J. A. (1949). The parasites and predators of potato aphids. *Bull ent. Res.* **40,** 97–122.

Sundby, R. A. (1966). A comparative study of the efficiency of three predatory insects *Coccinella septempunctata* L. (Coleoptera, Coccinellidae), *Chrysopa carnea* ST. (Neuroptera, Chrysopidae) and *Syrphus ribesii* L. (Diptera, Syrphidae) at two different temperatures. *Entomophaga* **11,** 395–404.

Withycombe, C. L. (1924). Note on the economic value of the Neuroptera, with special reference to the Coniopterygidae. *Ann. appl. Biol.* **11,** 112–25.

BEETLES

Beetles belong to the order Coleoptera; some 3,700 species are known in Britain. In the adult stage they are characterized by the modification of their fore wings into tough and horny veinless covers which protect the membraneous hind wings during rest and which are held erect and motionless in flight. In some beetles the hind wings are small or even absent, as in many ground beetles and certain weevils. They are strongly built insects with well-developed biting jaws. They have adopted all modes of life on land, in air and water to a greater extent than other insects, and have become the most numerous of all animals. It is not surprising that many have followed the predatory way of life; only the better-known kinds can be described here. Like the lacewings they have egg, larval and pupal stages before becoming adult beetles.

Ladybirds

Ladybirds are amongst the most familiar of beetles and almost all their common names in Europe as well as in this country, such as lady-cows, God's cows and the Virgin's insects, suggest an affectionate or at least respectful attitude to them by the public. John Curtis, a famous 19th century entomologist, remarked that they are among 'the few insects which in this country we are taught in our infancy not to regard with disgust or horror'. They form the family Coccinellidae with a little over 40 species in Britain. Except for one species (*Subcoccinella 24-punctata* (L.)) which feeds on leguminous plants and has occasionally occurred as a minor pest of crops including carnation, all are predatory both in their adult and larval stages, feeding on aphids, scale insects, mealybugs, thrips and mites. While a few are specific in their choice of food, many will eat a variety of prey. They are small or medium-sized beetles, oval or rounded in shape and very convex, often with bright shining colours and patterns of spots or patches on a contrasting background. They appear to be distasteful to birds and their conspicuous appearance is an example of warning coloration.

A common species is the two spot ladybird (*Adalia bipunctata* (L.)), Illustrated on Plate III, and though this typically has red wing-cases marked with two black spots, it is very variable and forms with black wing-cases and four or six red spots are not infrequent. The seven spot ladybird (*Coccinella septempunctata* L.), with red wing-cases marked with a total of seven black spots, is constant in its markings. Another common species is the ten spot ladybird (*Adalia decempunctata* (L.)), also red and black and very variable. *Thea 22-punctata* (L.) is yellow with 22 well-defined black spots; the 14 spot ladybird, *Propylea 14-punctata* (L.), (Plate III) is also bright yellow but with variable black markings. Other species are tawny, brownish or black, with or without red or yellow spots or patches according to species. A much less familiar one is the very small *Stethorus punctillum* Weise, which is a predator of fruit tree red spider mite and will be referred to in more detail later (page 20). Species of *Scymnus* are mostly sombrely coloured and attack aphids infesting coniferous trees; their larval stages are unusual in being covered with a waxy secretion so that they resemble the species of adelges on which they feed.

Ladybirds pass the winter as adults, sheltering under loose bark, in crevices in trees and walls, in the base of dense plant growth including grass, in hollow plant stems, and sometimes in houses. At hibernation time many species are gregarious, collecting in their winter quarters in numbers ranging from a few dozen to several hundred or even thousands. Often the same winter quarters are used year after year, and it has been suggested that the beetles are attracted by the characteristic odour of their kind, possibly including that of dead individuals remaining in the shelter. It has also been suggested that the value of these assemblages is the deterrent effect of the massed warning coloration. The beetles leave their hibernation quarters in the spring and then become solitary insects, any aggregations being due to chance such as the presence of suitable food supplies (Banks, 1954a). The adults of some species are known to live for more than one season.

In the spring the beetles pair and the females lay their eggs in clusters, rows or singly on the leaves and stems of plants and sometimes on stones or other non-living objects. The eggs are usually elongate, bright yellow or orange-yellow, and are deposited standing on end (Plate III). They hatch in two to six or eight days. In laboratory experiments (Sundby, 1966) with the seven spot ladybird, which has only one generation per year, the average number of eggs laid per female at 21°C (70°F) was 814, and eggs were laid over periods ranging from three days to 15 weeks. The average hatch was only 45 per cent. An average of over 1,500 eggs each were laid by some two spot ladybirds kept in captivity and fed on pea aphids (*Acyrthosiphon pisum* (Harris)).

The larvae or grubs, often called 'niggers', are of the general shape shown in Plate III and are active insects with six well-developed legs and a pair of strong jaws. They are often conspicuously coloured slaty blue with orange or yellowish spots and the body dotted with dark tubercles. The larvae moult three times and are fully grown in about three weeks. Their prey includes aphids (see page 16), mealybugs, scale insects, small caterpillars, and mites (see page 20). The fully grown larvae attach themselves to a leaf or some support by their rear ends and form the pupal stage, with the cast-off last larval skin frequently still attached. The pupa is plump, tapering to the attached extremity of the abdomen; the wings, legs and other appendages of the forthcoming adult beetle can be seen closely fixed to the body. The pupal skin eventually splits down the back to allow the adult to emerge. In many of the common species the pupa (Plate III) is marked with yellow and black, and especially when found on potato leaves is frequently mistaken for the Colorado beetle (*Leptinotarsa decemlineata* (Say).

Some half dozen of the British ladybirds are known to show mass movements and sudden appearances in large numbers, especially the two spot, seven spot and 11 spot (*Coccinella 11-punctata* L.) species. The reare many records of these sudden appearances, especially on the south and east coasts (Williams, 1958). In 1847 the cliffs of north-east Kent were swarming with at least three species of ladybird for a period of several days. In June 1925 there was an extensive eastward movement of two spot ladybirds across northern England. Especially large immigrations of the seven spot (also a great migrant as far abroad as India) were recorded in 1750, 1826 and 1847. In July and August 1952, along the south Lincolnshire coast, the high tide line was reddened by millions of beetles for a distance of 40 miles. The 11 spot ladybird (Plate III) is often associated with migrant swarms of the seven spot. Swarms of the seven spot were common in northern England in 1945, occurring on aphid-infested barley and turnips in late July, and larvae and pupae were unusually numerous on potato crops in August (Dunn, 1949). Apart from these sporadic mass movements, it may be that ladybirds are less abundant than they have been in the past, when John Curtis (1857) wrote of the 'congregation of myriads' sometimes occurring in hop gardens, and their often seen 'scarlet jackets sparkling on the bright green leaves' of turnip and corn fields.

Ladybirds and Aphids

As ladybirds are such common predators considerable attention has been given to various aspects of their behaviour. In a study of the suitability of various aphid species as food for ladybirds (Blackman, 1965, 1967a, 1967b), larvae of the two spot ladybird grew better, i.e., they grew more rapidly, a higher proportion survived and they produced heavier adults, when fed on nettle aphid (*Microlophium evansi* (Theob.)), pea aphid, peach-potato aphid and the mottled arum aphid (*Aulacorthum circumflexum* (Buckt.)) than when reared on the black bean aphid or the elder aphid (*Aphis sambuci* L.). Not only was black bean aphid less nutritious, but the larvae ate a smaller proportion of each aphid tackled than they did with the other species. The fecundity of the adults was, however, not affected by the diet of their larvae, but adult ladybirds fed on black bean aphid laid less than half the number of fertile eggs laid by adults which had been fed on peach-potato aphid. The vetch aphid (*Megoura viciae* Buckt.), which also occurs at times on peas and beans, was found to be lethal to both larvae and adults, though it was not refused when offered together with pea aphid. With the seven spot ladybird responses were different, for the larvae developed as well on a diet of black bean aphid as they did on peach-potato aphid or pea aphid. They could also grow to maturity on vetch aphid and cabbage aphid (*Brevicoryne brassicae* (L.)) though they did so more slowly.

Larvae of the ten spot ladybird also appear to find vetch aphid poisonous, and they have frequently been seen to regurgitate after eating black bean aphid (Dixon, 1958). The mealy plum aphid (*Hyalopterus pruni* (Geoff.))

is rejected by the larvae immediately after they have pierced one, and in subsequent encounters they will reject this aphid after touching it with their palps.

In laboratory studies on the feeding behaviour of the larvae (Blackman, 1967a), it was found that first stage larvae of the two spot ladybird rarely made a capture at their first encounter with an aphid and generally took several minutes to achieve success. Pea aphid with its rapid escape responses gave them most difficulty; in half the observed captures the aphid was seized by one of the terminal leg joints and managed to drag its attacker around for some time before it was eventually gripped by the body and overcome. The larvae took about two hours to feed on a peach-potato aphid or a pea aphid, eating most of the prey; they spent about six hours on a black bean aphid and then ate relatively little. Vetch aphid was also difficult to catch but was soon released, though even this brief feeding resulted in the death of the larvae. Older (fourth stage) larvae of two spot ladybird were more active than the first stage larvae, catching aphids more quickly. They spent a similar time feeding on, and ate similar proportions of, peach-potato aphid and pea aphid, but spent about five times as long on black bean aphid and ate less of each capture. They also readily attacked vetch aphid but after about four minutes suddenly stopped feeding and vomited.

The first stage larvae of the larger seven spot ladybird were more efficient, usually making a successful capture the first time they found an aphid. They spent similar times to those of first stage two spot ladybird larvae feeding on peach-potato aphid and pea aphid, and like them took longer over black bean aphid. They ate similar amounts of peach-potato aphid and black bean aphid, but more of pea aphid which seemed to be the most suitable prey. They took longer over vetch aphid and ate less of it, but were not poisoned. The fourth stage larvae proved to be very efficient, quickly capturing and eating all the four species of aphid offered them. They consumed their prey completely, except that one or two legs of pea aphid were always left and usually most of the skin of vetch aphid.

An interesting description of a ladybird larva tackling a winged aphid is given by John Curtis (1857). The 'little insect alligator', as he called it, threw its fore legs about the aphid, then gradually slid along to the wings which it bit until they were in a condition useless for flight. It then moved forward again, seized the aphid's thorax and ate into its side, at the same time holding down its struggling victim with one of its hind legs.

The number of peach-potato aphids eaten by larvae of the seven spot ladybird in laboratory experiments averaged 400 in 35 days at 15·5°C (60°F) (Sundby, 1966), and last stage larvae ate about ten black bean aphids per day (Banks, 1962).

Aphids are by no means passive victims (Dixon, 1958). When a ladybird larva seizes a black bean aphid, the other members of the colony set up rhythmic kicking movements. Small aphids are unable to resist being dragged away, but large aphids are sometimes able to kick away their attacker, or if seized manage to crawl away dragging the larva with them, and even free themselves and resume feeding although injured. Nettle aphids frequently avoid ladybird larvae by walking away, dropping from the plant, or kicking. If seized they may daub their attacker's head with a waxy secretion from their cornicles; this temporarily paralyses the larva, and if the aphid is held by a leg it may shake itself free but if its body is gripped it is unlikely to escape. Another species of aphid copiously waxes attacking predators to such an extent that some of them eventually die, while another avoids capture by jumping.

The two spot ladybird, one of the commonest predators on beans, is evidently not feeding on an aphid diet which will enable it to realize its full reproductive potential. The adults when laying their eggs do not appear to be influenced by the species of aphid present, though they may show some choice of habitat. For example, in experiments (Blackman, 1967b) with broad bean plants suspended at different heights, it was found that the seven spot ladybird laid most of its eggs near ground level, while the two spot ladybird preferred plants higher above ground. There is other evidence suggesting that vegetation height plays a part in habitat selection by ladybirds. It may be worth while to discover which ladybird species are the more effective predators and which wild habitat conditions most favour their successful breeding, so that adequate supplies are available to migrate to crops.

The course of events in the colonization by ladybirds of beans infested with black bean aphid has been studied at Rothamsted Experimental Station (Banks, 1955, 1956a). Regular observations were made on four plots of beans and three clumps of stinging nettles at various points on the farm at Rothamsted. The two spot ladybird was the most abundant, followed by the seven and 14 spot ladybirds. The ten and 11 spot ladybirds also occurred but were not common.

Ladybirds leaving hibernation paired and dispersed, many finding their way to the nettles infested with nettle aphid. Here eggs were laid and hatched to produce the 'nettle brood'. Some of the parent beetles died, others dispersed again and accumulated on the beans in the early stages of their infestation by black bean aphid. Here they laid more eggs to produce the 'bean brood'. Before the latter reached maturity, adults from the 'nettle brood' arrived on the beans at about the height of the aphid infestations. Finally, new adults from the 'bean brood' laid their eggs, giving rise to larvae when the aphid infestations were nearly ended; many of these larvae probably died of starvation. The sizes and times of rise and fall of the aphid and ladybird populations on the bean plots differed and were not closely synchronized. The number of beetles entering the plots was not obviously dependent on the sizes of the aphid infestations and seemed to be determined more by proximity of nettle beds which served as sources of beetles. The beetles tended to accumulate on heavily infested plants because of the plentiful food supply; they did not need the presence of aphids to stimulate them to lay eggs.

A suggestion of haphazardness in the behaviour of the adults is reflected in that of the larvae when searching for food. They are unable to sense an aphid until they touch it and, in a study of their behaviour on bean leaves (Banks, 1954b, 1957), they moved with frequent changes of direction, with a marked tendency to crawl along the edges of leaves and prominent veins. Much time and energy was often wasted in going over leaves already searched. On plants with few aphids the larvae could by chance move far away from them; one larva wandered for four days before coming within a few inches of an aphid colony and then did not find it. The larvae could travel as far as 21 ft in two days. When a larva found an aphid colony its behaviour changed: after feeding it made small turning movements from side to side, so increasing its chances of finding more aphids.

On plants with few aphids, many larvae, especially newly hatched ones, are likely to die of starvation because of their thorough but inefficient method of searching. This is to some extent offset by cannibalism, for newly hatched larvae were found to eat eggs from the batch from which they had hatched (Banks, 1956b). With the large or medium-sized egg clusters of the two and seven spot ladybirds, some 10–12 per cent of the fertile eggs were eaten by larvae already hatched, while with the small egg batches of the 14 spot ladybird the percentage eaten was about six. Infertile eggs were not common with the two and 14 spot ladybirds, but occurred regularly with the seven spot and were always eaten. This cannibalism had the advantage that consumption of one egg almost doubled the life of a larva and thus its opportunity of searching for aphids, while larvae which fed on two or three eggs lived even longer and could moult into the second stage larva without eating aphids. Cannibalism was believed to be responsible for a 56 per cent mortality of newly hatched young which occurred in one week on one of the Rothamsted bean plots.

An interesting observation was the effect of common black ants on the aphid infestations on beans: not only did the ants drive away ladybird larvae, but aphid colonies attended by ants multiplied slightly more rapidly than those free of ants (Banks, 1958, 1962). With comparatively light infestations of aphids—about 50 per stem—on plants in cages, seven spot ladybird larvae in their last stage killed about two-thirds of the aphids in two days on the plants to which ants had no access. The only aphids surviving were small ones in crevices among the topmost leaflets, where the predators could not reach them. In the cages with ants the ladybird larvae were immediately driven off and again attacked whenever they returned to the plants. Larvae of the two spot ladybird killed 70–90 per cent of the aphids in five days in the absence of ants, and again where ants were present the aphids were completely protected and their numbers increased geometrically. Ants have also been seen to drive away adult ladybird beetles from aphid colonies. In defence, both the adults and larvae discharge a yellow fluid which is repugnant to ants, but they still beat a retreat when threatened. Ants were also found to remove the ladybird eggs.

In a survey (Dunn, 1949) of the predators and parasites of aphids on potatoes in northern England in 1944 and 1945, the seven and two spot ladybirds were found, but were not usually very common and, except for odd occasions when the seven spot appeared in swarms, were probably only a contributory factor in the natural control of the aphids.

A study of the effects of temperature on ladybird predation of the pea aphid (Dunn, 1952) showed that at 5°C (41°F) the beetles were inactive while the aphids bred slowly; between 5°C and about 10·5°C (51°F) the aphids produced young faster than the beetles could consume them, while above 11°C (52°F) the beetles' consumption rate exceeded that at which the aphids could reproduce. Thus cool spring and autumn weather favours the aphid, while warm sunny weather increases the effectiveness of the predator.

Ladybirds and Spider Mites

The very small, black *Stethorus punctillum* (Plate III) is almost spherical in shape, only $\frac{1}{25}$ in. (1 mm) in diameter, with a covering of fine yellow hairs. The adults hibernate under bark and in other suitable places, emerging during May. The eggs, oval and pale yellow, are laid flat on the underside of leaves in May and June. The larvae, which are hairy, brown and flattened, crawl slowly among the leaf hairs in search of mites.

The larvae have three moults and when fully fed are $\frac{1}{10}$ in. (2·5 mm) long, changing into a black pupa attached by the tip of its abdomen to a leaf, usually on the underside. There are two generations a year, a further brood of beetles emerging in June and July and laying eggs in July and August. The adults live six to eight weeks and generally enter hibernation early in October. Both adults and larvae feed entirely on mites, principally the fruit tree red spider mite, with a preference for the larger stages, but its eggs are also attacked, especially by the larvae, and many of the winter eggs are devoured in the autumn. The adults consume the mites completely, with a daily average of 20 adult females, whereas the larvae suck mites and eggs until empty, accounting for an average of 24 mites a day (Collyer, 1953a).

This ladybird was found to be rare in a neglected Essex orchard studied by Collyer (1953a), possibly because its larvae cannot survive competition with other predatory insects, but was abundant in commercial orchards. In the latter, which were generally sprayed with an oil wash in the winter, it was usually scarce in the early part of the year but regularly recolonized the orchards from neighbouring hedgerows later in the summer, especially when the fruit tree red spider mite was abundant. It was capable of reducing the mite infestation in August, thereby leading to a smaller number of winter eggs being laid by the mite.

Pullus (=*Scymnus*) *auritus* (Thunb.), another small ladybird, is included in the list of 45 predators of the fruit tree red spider mite (Collyer, 1953a). The two spot and seven spot ladybirds also feed on these mites if aphids are scarce (Collyer, 1953b).

Stethorus punctillum also preys on red spider mite (*Tetranychus urticae* (Koch)) and was at one time common in hop gardens (Jary, 1935). It has been recorded as not uncommon on peach trees grown under glass in Holland, especially at the end of the season (Bravenboer, 1959). In these glasshouses it hibernates as an adult under the bark of the peach trees near soil level. The females lay an average of 4·6 eggs a day, with an average total of about 100, and they stop egg laying when the food supply becomes short. The period from egg to adult takes rather longer for the ladybird than for the red spider mite, being 20 days compared with 14, at 22°C (72°F). The beetles appear to prefer mites as food; they will eat aphids and whiteflies but do not thrive and will not lay eggs. The adults require 20–40 mites per day for subsistence, and may consume as many as 140 per day. The larvae, which do not chew but suck their prey, may account for 250 mites a day. It is calculated that starting with one female ladybird and 100 female red spider mites the latter would be eliminated after three generations. This predator showed promise as a biological aid in the control of red pider mite, and results of successful trials are described on page 107.

GROUND BEETLES

Ground beetles (family Carabidae) include nearly 350 species in Britain and most of them are sombrely coloured beetles living, as their name implies, in the soil and in damp, sheltered situations. Some are unable to fly but all are active creatures with long, slender legs and powerful, sharply pointed jaws. A few are pests, such as the strawberry seed beetle, *Harpalus rufipes* (Deg.), which sometimes damages strawberries by biting at the seeds and surrounding flesh, and the strawberry ground beetles, *Pterostichus madidus* (=*Feronia madida*) (F.) and *P. melanarius* (=*F. melanaria*) (Illig.) (=*P. vulgaris* (L.)), which occasionally eat cavities in the fruit.

Comparatively little is known about the life history and habits of the majority of ground beetles but, while many of them and their larvae are probably omnivorous or even live on vegetable matter to a greater extent than is generally realized, there is no doubt that the predatory habit is prominent in their activities. With the exception of those mentioned as damaging strawberries, they can be regarded as beneficial. Even the strawberry pests are only a nuisance when their numbers become unusually large; otherwise they are useful, particularly *Pterostichus* (=*Feronia*) species which are also carnivorous.

The larvae of ground beetles (Plate II) are well-armoured agile insects with comparatively narrow bodies, three pairs of slender legs, well-developed jaws and antennae, and a pair of feelers (cerci) on the last segment of the abdomen. During summer, eggs are laid in the soil where the larvae live; probably the adult stage of many species is reached in the following summer. Adults are, however, also to be found during the winter. They do not like strong sunlight and hunt mainly at night, especially in damp conditions when their larvae are probably also active on the surface of the soil. The beetles abhor drought and will drink greedily after being deprived of moisture.

Some of the larger species, such as *Carabus nemoralis* Müll. which has metallic bronze and green reflections, may be familiar to gardeners. The violet ground beetle (*C. violaceus* L.), which has violet margins to its wing-cases, is illustrated on Plate II. *Calosama inquisitor* (L.) lives in oak woods and is most frequent in the New Forest and Forest of Dean, where it climbs the trees to catch caterpillars, particularly those of the green oak tortrix moth and the winter moth (*Operophtera brumata* (L.)). It is a large beetle with a dark green or violet thorax and rich golden-green wing-cases with coppery reflections; its larva is black. The related *Calosoma sycophanta* (L.), common on the Continent, was imported into north-western United States of America during the introduction of natural enemies of the gipsy moth (*Lymantria dispar* (L.)), becoming established as a useful member of the North American fauna.

Ground Beetles and Cabbage Root Fly

Studies at the National Vegetable Research Station at Wellesbourne have shown that several species of predatory ground beetle and predatory rove beetle (see page 24) provide an important natural check of cabbage root fly (*Erioischia brassicae* (Bouché)). The natural mortality of this pest was found to be very high: during the period 1954–58 at Wellesbourne an average of more than 95 per cent of the eggs of the first and second generations in the spring failed to produce adult flies (Hughes and Salter, 1959). Mortality in the egg stage seemed to be the most important factor, and in 1957 and 1958 over 90 per cent of this mortality was accounted for by beetle predation (Hughes, 1959). An inverse relationship was found between egg survival and the numbers of the ground beetle *Bembidion lampros* (Herbst) (Plate III) caught in pitfall traps placed about four inches from the plants in plots of cauliflower (Wright, Hughes and Worrall, 1960).

Some 60 species of ground beetle and rove beetle were found in these traps, and investigations were made to find their relative importance as predators (Coaker and Williams, 1963a). To discover which species had been feeding on cabbage root fly, a serological technique was used. Rabbits were injected with a preparation of cabbage root fly proteins to induce formation of anti-bodies in their blood. The gut contents of the beetles could then be tested against serum obtained from the rabbits; the formation of a white precipitate produced by the combination of antigen from cabbage root fly and anti-body from the rabbit indicated that the beetle being examined had fed on one or other stages of the fly.

Of 30 species of ground beetle recorded in the plots, 13 were found to have fed on at least one stage of cabbage root fly. Six of these were the more important by virtue of their greater numbers, which together constituted over 90 per cent of the total for the 13 species. These six were *Bembidion lampros*, *B. quadrimaculatum* (L.), *Harpalus aeneus* (F.) (Plate III), strawberry seed beetle, *Pterostichus melanarius* (Plate II) and *Nebria brevicollis* (F.). The numbers of these beetles trapped from 1960–62 varied each year, but *B. lampros* and *H. aeneus* were regularly predominant from April to June, their numbers diminishing in July, while *P. melanarius* and strawberry seed beetle tended to become numerous in July when increased plant growth offered a habitat more favourable to them. Further study (Mitchell, 1963b), with pitfall traps placed directly under cabbage plants as well as between them, showed that species of *Trechus* were also important; *T. quadristriatus* (Schrank) occurred as numerously as *B. lampros*, though later in the season as the plants grew larger.

Observations of marked beetles (Coaker and Williams, 1963a) suggested that little or no emigration from the experimental area occurred, which was in agreement with other work (Thiele, 1960), indicating little intermixing of ground beetle populations of hedgerows and adjoining cultivated land. The population of ground beetles on the plots was estimated to be equivalent to about 6,000 per acre at any one time during April and May, and some 12 per cent of the beetles had fed on cabbage root fly in each 24 hour period with an average daily meal of about 18 root fly eggs.

The predatory values of some of the ground beetles were compared in laboratory experiments (Coaker and Williams, 1963a) in which beetles were enclosed with brassica plants and known numbers of cabbage root fly eggs. In three days *B. lampros* consumed an average of 73 per cent of the eggs. Using *B. lampros* as a standard and calling its predator value 1, the relative values for other species were: *B. quadrimaculatum* and *H. aeneus*, about 0·7; *N. brevicollis*, 0·4; strawberry seed beetle and *P. melanarius*, 0·23. *Trechus obtusus* Erich. had a value equal to that of *B. lampros*. Calculations suggested an average daily meal of 16 eggs per beetle. The most active beetles had the highest predator values, probably because eggs were largely found by chance contact. In earlier laboratory studies (Wright, Howells and Bowe, 1956) one individual of *B. lampros* ate 428 eggs, 159 first stage and three third stage larvae over a period of six months, and one individual of a *Pterostichus* species ate 180 pupae and 23 third stage larvae over a similar period.

The life history and habits of the two small ground beetles occurring in the largest numbers in the plots at Wellesbourne, *B. lampros* and *T. quadristriatus*, were investigated in more detail (Mitchell, 1963a, 1963b). Both these highly mobile and aggressive predators occur as short-winged (flightless) and normal long-winged forms; only the short-winged form of *B. lampros* and the long-winged form of *T. quadristriatus* were found at Wellesbourne. The two species were found to be mainly active by day, their activity depending upon air temperature; below about 9–10°C (48–50°F) *B. lampros* becomes inactive, while the threshold for activity with *T. quadristriatus* is below 4°C (40°F). Differences were found between the two species in the kind of habitat they prefer. *B. lampros* seeks bare ground, which would account for its abundance in the pitfall traps in the spring and decline in numbers as the brassica plants increase in size. During the day-time *T. quadristriatus* prefers the shaded conditions beneath the plants, its numbers increasing as the size of the plants increases.

B. lampros lays its eggs in the spring, the larvae which hatch maturing and producing a new generation of adults later in the summer. *T. quadristriatus* breeds in the autumn, its larvae overwintering and becoming adult beetles in the following spring; a small generation of eggs and larvae in the spring may be a continuation of the main autumn brood. The adults of both species can survive for at least a year. Adults and larvae of the two species appear to feed entirely on animal prey and animal debris. The adults mainly attack small insects, mites and related creatures active on the soil surface, though they will also feed on immobile animal matter such as dead insects and the eggs of cabbage root fly. They hunt with jaws held open, recognizing live prey by sight and finding other animal matter by chance physical contact. Movement of nearby prey, even one of their own species, stimulates them to attack. Having found and eaten a meal, their behaviour changes for a few moments: they walk slowly and make short turning movements, so that neighbouring food material is found more readily than it would be by random searching. Beetle numbers were found to increase with increasing numbers of cabbage root fly eggs and to decline as the latter decreased. The larvae, which also move about with open jaws, recognize prey by contact and feed on slow-moving, soft-bodied animals—probably mainly small earthworms and eelworms—and animal remains found in the soil.

The importance of ground beetles in relation to control of cabbage root fly, the use of insecticides and the development of resistance to them by the fly, are discussed on pages 79–81.

Ground Beetles and Other Pests

In laboratory studies (Stephenson, 1965) in which seven species of ground beetle were compared for their ability to prey on the field slug (*Agriolimax reticulatus* (Müll.)), the violet ground beetle was the most voracious and ate five slugs in four days. *Pterostichus melanarius* ate five slugs in 14 days, and *Abax parallelopipedus* (Pill. & Mitt.) four in ten days. *P. niger* (Schall.) and *P. madidus* ate few slugs, and the strawberry seed beetle none at all.

Some of the ground beetles, including species of *Harpalus*, *Pterostichus* and especially *Trechus quadristriatus* and *T. obtusus*, are suspected of being important predators of lettuce root aphid (Dunn, 1960), and certain species also eat the eggs of wheat bulb fly (*Leptohylemyia coarctata* (Fall.)) (Ryan, 1965, 1966).

Rove Beetles

The rove beetles (Staphylinidae) comprise another large family of over 950 British species which live in soil, decaying vegetable and animal matter, etc., and include many species that are predacious. They have rather narrowed bodies with shortened wing-cases leaving part of the abdomen exposed. The devil's coach-horse, *Staphylinus olens* Müll., is probably the only species that is generally familiar (Plate III). Rove beetles have active larvae which are similar to those of ground beetles, and many are known to be predatory (Plate III).

The investigation at the National Vegetable Research Station into the effects of predatory beetles on the survival of the young stages of cabbage root fly (Coaker and Williams, 1963a) showed that rove beetles were important as well as ground beetles. By using the serological test described on page 22, it was found that at least nine species were feeding on immature stages of the fly. The most common species in order of importance were *Aleochara bilineata* Gyll., *A. bipustulata* (L.) (Plate III) and *Oxytelus rugosus* (F.). The other species were *Philonthus succiola* Thom., *P. cognatus* Steph., *Tachyporus hypnorum* (F.), *T. solutus* Erich., *T. obtusus* (L.) and *Xantholinus longiventris* Heer. Average numbers in the soil round brassica plants in experimental plots varied from 2–6 per plant, and population estimates in May 1961 suggested that rove beetles were twice as numerous as the ground beetles. The larvae as well as the adults of these species are predatory, attacking the eggs and larvae of cabbage root fly. The pupal stage of the fly is also attacked, the larvae of *Aleochara* species living as parasites in the puparia. Up to 65 per cent of puparia have been recorded as parasitized by *Aleochara* species and the cynipid wasp *Idiomorpha rapae* (Westw.) (Hughes, Salter and Mitchell, 1959).

Aleochara species parasitize the pupae of wheat bulb fly and are being studied together with other parasites, predators and diseases which cause an appreciable mortality of the young stages of this pest (Ryan, 1965). They are also among the parasites of carrot fly (*Psila rosae* (F.)) (Wright, Geering and Ashby, 1947).

Species of *Atheta*, *Aleochara*, *Oxytelus* and *Tachyporus* appear likely to be important predators of lettuce root aphid (Dunn, 1960). Larvae of *T. obtusus* and *T. hypnorum* were recorded as preying upon strawberry aphid (*Chaetosiphon fragaefolii* (Cock.)) and occurring wherever aphids were numerous on strawberry crops examined in a survey in south-east England in 1943 (Dicker, 1944). They were plentiful in May and June, with small numbers persisting until early August, though they were not able to make much impression on the aphids which greatly outnumbered them.

Oligota flavicornis Boisd., a very small, shining, black rove beetle, about

$\frac{1}{25}$ in. (1 mm) in length, is a predator of the fruit tree red spider mite (Collyer, 1953a). It overwinters under bark, in vegetation, etc., becoming active in May and laying eggs in May and June; a second generation of beetles lays eggs in July and August. The eggs are rounded or oval, shining and almost colourless until becoming dark orange before hatching, and are laid on the lower surface of leaves and covered with a pile of debris such as cast or sucked mite skins, leaf hairs, etc. They hatch in four to seven days, the cycle of egg to adult occupying about four weeks. The larvae are pale yellow and narrow-bodied with three pairs of legs. The adult beetles completely devour the mites, while the larvae suck them, and they will suck winter eggs readily in September if mites become scarce; neither have been seen to take other prey. In the study begun in 1945 of the predators of this pest, the rove beetle was sometimes found to be numerous in Essex orchards when fruit tree red spider mite infestations were severe.

Other Predatory Beetles

Over 40 families of beetles include predatory species. The tiger beetles (subfamily Cicindelinae) are fiercely carnivorous, usually found on sandy heaths and moorlands where their larvae dig tunnels in which to lurk and await passing victims. The adults fly in sunshine and are extremely active and formidable hunters of other insects. Larvae and adults of water beetles such as *Dytiscus* species (family Dytiscidae) are voracious predators in still water, attacking small fish as well as insects such as mosquito larvae. *Xylodrepa quadripunctata* (L.) (family Silphidae, carrion beetles) is common in oak woods and preys on moth caterpillars, especially those of the winter moth and mottled umber moth (*Erannis defoliaria* (Clerck)). The black larvae, suggestive of elongated woodlice in shape, live in the soil.

The glow-worm (*Lampyris noctiluca* (L.) (family Lampyridae)) is the larval stage of a beetle that feeds upon snails and slugs. The prey is seized by the larva's sickle-shaped jaws, down which passes a secretion that predigests the tissues of the prey; the larva then imbibes the resulting fluids. The bright blue beetle, *Korynetes caeruleus* (Deg.) (family Cleridae), preys upon furniture beetles (*Anobium* species) and the death watch beetle (*Xestobium rufovillosum* (Deg.)). In pine woods the ant beetle, *Thanasimus formicarius* (L.) (Cleridae), hunts the tree trunks for the injurious bark beetles, and its red larva feeds on the bark beetles' eggs and larvae in their galleries in the bark. Minute beetles of the family Nitidulidae are important predators of bark beetles.

References and Further Reading

Banks, C. J. (1954a). Random and non-random distributions of Coccinellidae. *J. Soc. Br. Ent.* **4,** 211–15.

Banks, C. J. (1954b). The searching behaviour of coccinellid larvae. *Br. J. Anim. Behav.* **2,** 37–38.

Banks, C. J. (1955). An ecological study of Coccinellidae (Col.) associated with *Aphis fabae* Scop. on *Vicia faba*. *Bull. ent. Res.* **46,** 561–87.

Banks, C. J. (1956a). The distributions of coccinellid egg batches and larvae in relation to numbers of *Aphis fabae* Scop. on *Vicia faba*. *Bull. ent. Res.* **47,** 47–56.

BANKS, C. J. (1956b). Observations on the behaviour and mortality in Coccinellidae before dispersal from the egg shells. *Proc. R. ent. Soc. Lond. (A)* **31,** 56–60.

BANKS, C. J. (1957). The behaviour of individual coccinellid larvae on plants. *Br. J. Anim. Behav.* **5,** 12–24.

BANKS, C. J. (1958). Effects of the ant, *Lasius niger* (L.), on the behaviour and reproduction of the black bean aphid, *Aphis fabae* Scop. *Bull. ent. Res.* **49,** 701–14.

BANKS, C. J. (1962). Effects of the ant *Lasius niger* (L.) on insects preying on small populations of *Aphis fabae* Scop. on bean plants. *Ann. appl. Biol.* **50,** 669–79.

BLACKMAN, R. L. (1965). Studies on specificity in Coccinellidae. *Ann. appl. Biol.* **56,** 336–8.

BLACKMAN, R. L. (1967a). The effects of different aphid foods on *Adalia bipunctata* L. and *Coccinella 7-punctata* L. *Ann. appl. Biol.* **59,** 207–19.

BLACKMAN, R. L. (1967b). Selection of aphid prey by *Adalia bipunctata* L. and *Coccinella 7-punctata* L. *Ann. appl. Biol.* **59,** 331–8.

BRAVENBOER, L. (1959). De Chemische en Biologische bestrijding van de spintmijt *Tetranychus urticae* Koch. *Publikatie van het Proefstation voor de Groenten-en Fruitteelt onder Glas te Naaldwijk*, No. 75.

CHRYSTAL, R. NEIL (1937). *Insects of the British woodlands.* London and New York, Frederick Warne.

CLAUSEN, C. P. (1940). *Entomophagous insects.* New York and London, McGraw-Hill.

COAKER, T. H. and FINCH, S. (1965). Natural control of the cabbage root fly. *Rep. natn. Veg. Res. Stn for 1964*, 60.

COAKER, T. H. and WILLIAMS, D. A. (1963a). The importance of some Carabidae and Staphylinidae as predators of the cabbage root fly, *Erioischia brassicae* (Bouché). *Entomologia exp. appl.* **6,** 156–64.

COAKER, T. H. and WILLIAMS, DOROTHY A. (1963b). The natural control of the cabbage root fly. *Rep. natn. Veg. Res. Stn for 1962*, 50.

COAKER, T. H. and WILLIAMS, DOROTHY A. (1964). Natural control of the cabbage root fly. *Rep. natn. Veg. Res. Stn for 1963*, 58.

COAKER, T. H. and WORRALL, JENNIFER (1961). The biology and natural control of the cabbage root fly. *Rep. natn. Veg. Res. Stn for 1960*, 42.

COLLYER, ELSIE (1953a). Biology of some predatory insects and mites associated with the fruit tree red spider mite (*Metatetranychus ulmi* (Koch)) in south-eastern England. II. Some important predators of the mite. *J. hort. Sci.* **28,** 85–97.

COLLYER, ELSIE (1953b). Biology of some predatory insects and mites associated with the fruit tree red spider mite (*Metatetranychus ulmi* (Koch)) in south-eastern England. III. Further predators of the mite. *J. hort. Sci.* **28,** 98–113.

CURTIS, JOHN (1857). *Farm insects.* London, Blackie and Son.

DICKER, G. H. L. (1944). *Tachyporus* (Col: Staphylinidae) larvae preying on Aphides. *Entomologist's mon. Mag.* **80,** 71.

DIXON, A. F. G. (1958). The escape responses shown by certain aphids to the presence of the coccinellid *Adalia decempunctata* (L.). *Trans. R. ent. Soc. Lond.* **110,** 319–34.

DUNN, J. A. (1949). The parasites and predators of potato aphids. *Bull. ent. Res.* **40**, 97–122.

DUNN, J. A. (1952). The effect of temperature on the pea aphid–ladybird relationship. *Rep. natn. Veg. Res. Stn for* 1951, 21–3.

DUNN, J. A. (1960). The natural enemies of the lettuce root aphid, *Pemphigus bursarius* (L.). *Bull. ent. Res.* **51**, 271–8.

EL HARIRI, G. (1966). Laboratory studies on the reproduction of *Adalia bipunctata* (Coleoptera, Coccinellidae). *Entomologia exp. appl.* **9**, 200–4.

HAGEN, K. S. (1962). Biology and ecology of predaceous Coccinellidae. *A. Rev. Ent.* **7**, 289–326.

HUGHES, R. D. (1959). The natural mortality of *Erioischia brassicae* (Bouché) (Diptera, Anthomyiidae) during the egg stage of the first generation. *J. Anim. Ecol.* **28**, 343–57.

HUGHES, R. D. and SALTER, DOROTHY D. (1959). Natural mortality of *Erioischia brassicae* (Bouché) (Diptera, Anthomyiidae) during the immature stages of the first generation. *J. Anim. Ecol.* **28**, 231–41.

HUGHES, R. D., SALTER, DOROTHY D. and MITCHELL, B. (1959). The natural control of cabbage root fly at Wellesbourne. *Rep. natn. Veg. Res. Stn for 1958*, 31.

JARY, S. G. (1935). Some observations upon the 'red spider', *Tetranychus telarius* L., on hops and its control, with notes on some predatory insects. *Ann. appl. Biol.* **22**, 538–48.

JONES, MARGARET G. (1967). Predators of frit fly. In: Johnson, C. G. Entomology department. *Rep. Rothamsted exp. Stn for 1966*, 203–4.

LINSSEN, E. F. (1959). *Beetles of the British Isles, Series I and II.* The Wayside and Woodland Series. London and New York, Frederick Warne.

MITCHELL, B. (1963a). Ecology of two carabid beetles, *Bembidion lampros* (Herbst) and *Trechus quadristriatus* (Schrank). I. Life cycles and feeding behaviour. *J. Anim. Ecol.* **32**, 289–99.

MITCHELL, B. (1963b). Ecology of two carabid beetles, *Bembidion lampros* (Herbst) and *Trechus quadristriatus* (Schrank). II. Studies on populations of adults in the field, with special reference to the technique of pitfall trapping. *J. Anim. Ecol.* **32**, 377–92.

RYAN, M. F. (1965). Wheat-bulb fly predators. In: Johnson, C. G. Entomology department. *Rep. Rothamsted exp. Stn for 1964*, 182.

RYAN, M. F. (1966). Predators of wheat-bulb fly. In: Johnson, C. G. Entomology department. *Rep. Rothamsted exp. Stn for 1965*, 194.

RYAN, M. F. (1967). Predation of wheat-bulb fly. In: Johnson, C. G. Entomology department. *Rep. Rothamsted exp. Stn for 1966*, 204–5.

STEPHENSON, J. W. (1965). Slug parasites and predators. In: Johnson, C. G. Entomology department. *Rep. Rothamsted exp. Stn for 1964*, 188.

SUNDBY, R. A. (1966). A comparative study of the efficiency of three predatory insects *Coccinella septempunctata* L. (Coleoptera, Coccinellidae), *Chrysopa carnea* ST. (Neuroptera, Chrysopidae) and *Syrphus ribesii* L. (Diptera, Syrphidae) at two different temperatures. *Entomophaga* **11**, 395–404.

THIELE, H. U. (1960). Gibt es Beziehungen zwischen der Tierwelt von Hecken und angrenzenden Kulturfeldern? *Z. angew. Ent.* **47**, 122–7.

WILLIAMS, C. B. (1958). *Insect migration.* New Naturalist Series, No. 36. London, Collins.

Wright, D. W., Bowe, Dorothy D. and Buxton, J. (1955). Cabbage root fly. *Rep. natn. Veg. Res. Stn for 1954*, 53.

Wright, D. W., Geering, Q. A. and Ashby, D. G. (1947). The insect parasites of the carrot fly, *Psila rosae* Fab. *Bull. ent. Res.* **37**, 507–29.

Wright, D. W., Howells, W. R. and Bowe, Dorothy D. (1956). Natural enemies of the cabbage root fly. *Rep. natn. Veg. Res. Stn for 1955*, 47.

Wright, D. W., Hughes, R. D., Salter, Dorothy D. and Worrall, Jennifer (1960). The biology and natural control of the cabbage root fly. *Rep. natn. Veg. Res. Stn for 1959*, 35.

Wright, D. W., Hughes, R. D. and Worrall, J. (1960). The effect of certain predators on the numbers of cabbage root fly (*Erioischia brassicae* (Bouché)) and on the subsequent damage caused by the pest. *Ann. appl. Biol.* **48**, 756–63.

PARASITIC WASPS

The important order Hymenoptera, with some 6,000 species in this country, embraces the sawflies, bees, wasps, ants and several large groups of parasitic insects. The adults have two pairs of membraneous wings; the narrower and shorter hind wings are provided on their front margin with a row of minute hooks which enable them to be linked with the corresponding fore wings. Except in the sawflies, the first segment of the abdomen is constricted to form a narrow waist. The insects pass through larval and pupal stages in their growth, the larvae being caterpillar-like in the sawflies and legless grubs in the other groups; in the parasitic groups the larvae are modified for their peculiar mode of life.

The order is subdivided into 21 major groups or superfamilies. One of these comprises the social and solitary bees which are important as pollinators and are referred to on pages 45–50. The superfamily which includes the social wasps also numbers many solitary wasps, smaller and more slender-bodied than the common wasp but frequently banded with black and yellow or orange, or coloured metallic blue, and there is a further superfamily of solitary insects which includes the digger wasps. Many of these wasps are predatory in habit, catching other insects which they sting into immobility to store in burrows as food for their grubs. Their prey includes caterpillars, weevils, flies, bees, froghoppers and spiders, most wasp species having a particular preference for one type of food.

The parasites belong to seven superfamilies containing many hundreds of species, some of which—the gall wasps—produce galls on plants. Many of the more important and familiar parasitic species belong to the families Ichneumonidae and Braconidae and the superfamily Chalcidoidea, representatives of which are shown in Plate IV. Most species place their eggs on or inside the body of the host insect by means of a long, needle-like ovipositor issuing from the abdomen of the female. This organ is composed of two halves, each semi-circular in cross-section, which when held together form a tube down which the egg is passed. The parasite larva usually lies free in the body fluid of the host insect, obtaining its oxygen from the latter's blood. Such larvae are generally very simple in form, with no eyes and no movable appendages except jaws. The insects that are parasitized may be in the larval, pupal or even the egg stage.

Ichneumons

The ichneumon flies (family Ichneumonidae), of which over 1,800 species have so far been identified in Britain, are for the most part parasites of the caterpillars of butterflies and moths. However, some parasitize their own order, especially sawflies, others favour beetles and, to a smaller extent, flies, while a few parasitize spiders, aphids and lacewings. It appears that most ichneumons do not usually confine themselves to any particular species of host. They are sun-loving insects, frequenting flowers, and show remarkable instincts in discovering their prey. The females of many species hibernate as adults, the males dying before winter.

The internally parasitic larvae have a distinct head and in many species possess a long tail-like process in their early stages of life. In some species the larvae are external parasites; they are generally hairy with a well-developed head but no tail process. Most of these attack insects in protected situations such as in rolled leaves and cocoons, or quiescent insects such as pupae, but some are adapted for life on more mobile hosts. For example, *Paniscus* species lay small, shining black eggs on caterpillars; these eggs are anchored by a short stalk which is embedded in the skin of the victim where it swells to make its hold more secure, and the larvae, when they hatch, remain adhering to the egg shell, feeding on the parts of the host in the vicinity (Plate IV). When ready for pupation, ichneumon larvae spin silken cocoons, often composed of iridescent black, white or yellow strands, either within the emptied body of the host or outside it.

The red ichneumons (*Ophion* species) (Plate IV) are attracted to light and are familiar insects; they are common parasites of moth caterpillars. The giant *Rhyssa persuasoria* (L.) parasitizes the wood-boring larvae of the wood wasps, showing the most remarkable powers of detecting the presence of these larvae deep in their tunnels in trees, and of insinuating their long, flexible and almost hair-like ovipositor through the wood, often for several inches, to reach them. Ichneumon larvae are often solitary in their hosts, the females apparently being able to recognize and avoid insects already attacked. Some, however, such as *Pimpla* species, attack parasite larvae and are therefore hyperparasites. Caterpillars of the diamond-back moth may be parasitized to the extent of 90 per cent of their numbers by *Angitia fenestralis* (Holmgr.) (Smith, 1931).

Braconids

The Braconidae, with nearly 900 British species, also parasitize a wide range of insects, more especially butterfly and moth caterpillars, though there are some important members attacking aphids. Most are internal parasites and the larvae often have a tail-like process similar to that found in ichneumons. Pupation may occur within the host's body but more usually outside it, and a cocoon is constructed. This may be of silk or, as in *Microgaster* species, of a shining paper-like material. The larvae of some species are gregarious and when they emerge from the host build honeycomb pattern. *Apanteles glomeratus* (L.) (Plate IV) is a common parasite of the caterpillars of the large white butterfly (*Pieris brassicae* (L.)), and the cluster of sulphur-yellow cocoons on the collapsed skin of the host is a familiar sight. One caterpillar may support as many as

150 parasite larvae. *Aphidius* species and related wasps attack aphids, usually wingless females. When one of these larvae is ready to pupate it cuts through the underside of the aphid's body and cements this to a leaf; the straw-coloured aphid skins, with a circular hole through which an adult parasite has emerged, may frequently be seen (Plate IV).

The primary parasite of cabbage aphid found in experimental plots studied in 1953 and 1954 (George, 1957) was *Diaeretiella* (= *Diaeretus*) *rapae* (Curt.), a parasitic wasp related to *Aphidius* species. This parasite also attacked the peach-potato aphid on caged plants, but not black bean aphid. It overwintered in mummified cabbage aphids, but quite high proportions sometimes failed to emerge the following summer. The peach-potato aphid is frequently parasitized by *Aphidius matricariae* Hal. In a survey of the parasites and predators of aphids on potato crops in northern England (Dunn, 1949), *A. avenae* Hal. and *Praon volucre* (Hal.) were found parasitizing the peach-potato aphid and two other common aphids on potatoes, the potato aphid (*Macrosiphum euphorbiae* (Thos.)) and the glasshouse-potato aphid (*Aulacorthum solani* (Kltb.)). *Aphidius avenae*, which attacks numerous other aphid species, was the commonest parasite found in this survey. Parasitized aphids first appeared about mid-July and there were probably two generations a year, the insects hibernating as larvae in mummified aphid skins. Numerous hyperparasites were also identified and appeared to be reducing the first generation of *A. avenae* by about one half.

The potential value of *A. matricariae* for the biological control of the peach-potato aphid on glasshouse chrysanthemums is being studied at the Glasshouse Crops Research Institute (Wyatt, 1966). Introduction of the parasites to chrysanthemum plots resulted in a 25 per cent reduction of the aphid population in five weeks. The successful control of red spider mites by an introduced predator (see page 94) and the difficulty of obtaining satisfactory chemical control of aphids have stimulated great interest in the possibilities of integrating biological and other control methods on glasshouse crops.

The braconid *Dacnusa gracilis* (Nees) was one of the two principal parasites of the carrot fly in a study made in East Anglia in 1944 and 1945 (Wright, Geering and Ashby, 1947). It parasitizes the larvae and pupae of carrot fly and has two generations a year coinciding with those of its host. The degree of parasitism of the first generation of carrot fly ranged from about 20 to nearly 50 per cent, and that of the overwintering generation was usually about 20 per cent. The second principal parasite was *Loxotropa tritoma* (Thoms.) (family Diapriidae), which has only one generation a year, the adults emerging towards the end of the first carrot fly generation. Like *D. gracilis* it was found to be widely distributed and sometimes parasitized 12–20 per cent of the larvae.

In a study of the relationship between the attendance of ants on aphids and predation by natural enemies (Banks, 1962), it was found that while ladybird, hover fly and lacewing larvae were driven off by ants, a parasitic wasp of *Diaeretus* species was not attacked and little disturbed by the ants while laying its eggs.

CHALCIDS

The term 'chalcid' refers to the group of families called the Chalcidoidea, which contains over 1,500 species in Britain; some of them are among the smallest known insects. The majority are either parasites or hyperparasites, the parasites being of even greater economic importance as enemies of insect pests than the ichneumons and braconids. The most commonly attacked insects are butterflies and moths, flies, scale insects and other Hemiptera, usually in the egg or larval stages, though some chalcids prefer larvae about to pupate or newly formed pupae.

As a rule no cocoon is formed and pupation occurs within or near the remains of the host. The curious phenomenon of polyembryony occurs in species of *Ageniaspis*, *Litomastix* and *Copidosoma* (all members of the family Encyrtidae). The females deposit their eggs in those of Lepidoptera; the larvae of the latter emerge in the normal way, but in the body of the larva the parasite's egg undergoes embryonic subdivision, producing a long string of embryos enclosed in a sheath. The number of embryos varies: in some species it is 150–200 but about 1,500 in *Litomastix truncatellus* (Dalman). *Ageniaspis fuscicollis* (Dalman) is a common parasite of the small ermine moths (*Yponomeuta* species).

Pteromalus puparum (L.) is one of the commonest chalcids, parasitizing especially the white butterflies. *Nasonia brevicornis* Ashm. parasitizes the pupae of the house fly, blowflies and related flies. *Aphelinus* species attack scale insects and aphids, and one species, together with *Encarsia formosa* Gahan, is referred to again on pages 98–100. Several other chalcids also parasitize aphids. *Blastothrix sericea* (Dalman) parasitizes scale insects (Plate IV). *Trichogramma* species favour the eggs of Lepidotera; as many as 20 individuals may develop in a single egg. Certain chalcids swim in water to attack water beetles, dragonflies and water boatmen. *Copidosoma tortricis* Wtrst., with a wing span of only 3 mm, is capable of giving an almost complete control of the strawberry tortrix moth (*Acleris comariana* (Zell.)), whose caterpillars sometimes infest strawberry leaves but are nearly always kept to low numbers by the parasite. As many as 35 parasites have been observed to emerge from one caterpillar (Massee, 1954).

Some hymenopterous parasites kill their hosts not only by developing in them but by the adults puncturing the host's body with their ovipositors and feeding on the body fluids exposed. This is known, for example, with the encyrtid *Aphycus* (=*Metaphycus*) *helvolus* Comp., which has been studied as a parasite of the scale insect *Saissetia oleae* (Bern.) on citrus trees in California. The parasite could destroy some 70–97 per cent of a scale insect population in a few months, about three-quarters of the mortality being due to feeding on the host (DeBach, 1943). Although this habit enables such a parasite to decrease pest numbers more rapidly, it also means that it might be less efficient where host numbers are low, since by depleting the supply of hosts by feeding on them, few individuals would be left to enable the parasite to reproduce itself (Flanders, 1953).

Other Parasitic Wasps

There are two further groups, the Proctotrupoidea and the Cynipoidea (gall wasps), which number over 1,000 species in Britain and include many parasites. The cynipid wasp *Idiomorpha rapae*, for example, is a parasite of cabbage root fly, laying its eggs on the first or second stage larvae; the adult emerges from the puparium. Together with rove beetles (see page 24) it has been recorded as affecting 65 per cent of the pupae of this pest (Hughes, Salter and Mitchell, 1959). Numerous species of these small wasps are recorded among the parasites of insect pests, though our knowledge of the extent to which they help to keep them in check is still comparatively limited.

References and Further Reading

BANKS, C. J. (1962). Effects of the ant *Lasius niger* (L.) on insects preying on small populations of *Aphis fabae* Scop. on bean plants. *Ann. appl. Biol.* **50,** 669–79.

CLAUSEN, C. P. (1940). *Entomophagous insects*. New York and London, McGraw-Hill.

DEBACH, PAUL (1943). The importance of host-feeding by adult parasites in the reduction of host populations. *J. econ. Ent.* **36,** 647–58.

DUNN, J. A. (1949). The parasites and predators of potato aphids. *Bull. ent. Res.* **40,** 97–122.

FLANDERS, S. E. (1953). Predatism by the adult hymenopterous parasite and its role in biological control. *J. econ. Ent.* **46,** 541–4.

GEORGE, K. S. (1957). Preliminary investigations on the biology and ecology of the parasites and predators of *Brevicoryne brassicae* (L.). *Bull. ent. Res.* **48,** 619–29.

HUGHES, R. D., SALTER, DOROTHY D. and MITCHELL, B. (1959). The natural control of cabbage root fly at Wellesbourne. *Rep. natn. Veg. Res. Stn for 1958*, 31.

MASSEE, A. M. (1954). *The pests of fruits and hops*. London, Crosby Lockwood.

SMITH, K. M. (1931). *A textbook of agricultural entomology*. Cambridge, University Press.

STEP, EDWARD (1932). *Bees, wasps, ants and allied insects of the British Isles*. The Wayside and Woodland Series. London and New York, Frederick Warne.

WRIGHT, D. W., GEERING, Q. A. and ASHBY, D. G. (1947). The insect parasites of the carrot fly, *Psila rosae* Fab. *Bull. ent. Res.* **37,** 507–29.

WYATT, I. J. (1966). *Myzus persicae* on chrysanthemums. *Rep. Glasshouse Crops Res. Inst. for 1965*, 88–9.

FLIES

The name 'fly' is used by entomologists to refer to members of the order Diptera. These insects when adult have a single pair of membraneous wings (a very few are wingless), the hind wings being represented by a pair of small drumstick-like organs called 'halteres'. The mouthparts are usually in the form of a short, soft proboscis for sucking liquid food such as nectar, but in blood-sucking flies such as mosquitoes, stable flies and horse flies they are stylet-like or spear-like for piercing and sucking. The growth stages of flies are egg, larva and pupa; the larva is characteristically legless and frequently appears to be headless as well. There are about 5,000 species in Britain; a considerable number are serious pests, but there are several groups which include important predatory and parasitic flies. Flies are active in sunshine, many frequenting flowers and acting as pollinating agents (see page 46).

Hover Flies

Hover flies belong to the family Syrphidae with nearly 250 species found in Britain, many of which are familiar through their habit of remaining poised in the air at one spot, then darting to another, their wings vibrating so rapidly as to be almost invisible. The flies, which are frequently banded with wasp-like black and yellow colours, are common in the summer and autumn and are often seen visiting flowers. A few are pests, e.g., the large narcissus fly, *Merodon equestris* (F.). The larvae of many live in decaying organic matter and some (*Volucella* species) as scavengers in ants' nests. The larvae of several common kinds, for example species of *Eristalis* and *Helophilus*, live in decaying animal and vegetable matter that is in a fluid condition; their bodies are terminated by a long slender breathing-tube which has earned them the name of 'rat-tailed maggots'.

The larvae of a considerable number of hover flies, however, feed on aphids, and in some seasons they are plentiful and make a considerable impression on aphid infestations. The minute cylindrical eggs, laid on leaves, are pale yellow or whitish and hatch in a few days. The larvae (Plate V) have soft bodies, broad at the tail end and tapering forwards, with no distinct head like that to be seen, for example, in larvae of beetles or moths. The head region is provided with a pair of mouth-hooks with which the insect seizes and lacerates its prey so that it can suck up the body contents. When feeding, the larva holds its victim aloft so that its struggles are of no avail. The body is whitish or lightly tinted, with a wrinkled skin which is semi-transparent so that movement of the internal organs can be seen. On the broad terminal segment of the abdomen is a short tube bearing a pair of breathing pores or spiracles. When fully fed, the larva attaches itself to a leaf, contracts and becomes still; its skin hardens to form a case known as the puparium (Plate V)—a feature characteristic of certain groups of the Diptera—within which pupation occurs. The adult fly emerges in due course; with some species in summer this may occur after only a few days. Some species appear to have but one generation a year while others have several. The insects hibernate usually as larvae or pupae.

In laboratory experiments (Sundby, 1966) with *Syrphus ribesii* (L.) the average number of eggs laid by a female at 21°C (70°F) was about 140, and the period from egg to adult occupied about three weeks. The average percentage hatch of the eggs was 67 and, with two generations in the year, it was calculated that the progeny produced during the summer from one female emerging in the spring would be nearly 4,000. Females of *S. luniger* (Meig.) began egg laying about the sixth week after emergence and laid an average of about 26 eggs each a day over a period of two weeks (Dixon, 1959). In the absence of aphids the females of this species were able to withhold egg laying for several weeks, which would enable them to survive periods of unfavourable weather or scarcity of aphids.

Hover fly larvae are voracious feeders. In some laboratory experiments (Sundby, 1966) larvae of *S. ribesii* consumed about 600 peach-potato aphids each in the 15 days of the larval stage at 15·5°C (60°F), but fewer aphids at 21°C (70°F). First stage larvae took about 80 minutes to eat one aphid, while fully grown larvae could eat one in about four minutes. Larvae of *S. balteatus* (Deg.) in captivity consumed 230–600 cabbage aphids each during their lifetimes (George, 1957). *Platycheirus manicatus* (Meig.) completed larval growth on 71 aphids from potatoes, and a larva of *S. vitripennis* (Meig.) consumed 112 (Dunn, 1949). One larva of *S. nitidicollis* (Meig.) has been recorded as eating 21 large aphids in 20 minutes, and one kept in captivity disposed of 800–900 aphids in its lifetime.

It was noticed in work on aphid predators that hover flies seemed to be less haphazard than ladybirds when laying their eggs, placing them more often close to aphid colonies, so that on hatching the larvae would find themselves near a food supply (Banks, 1953). The amount of discrimination shown by the females when choosing a place to lay eggs, and the stimuli causing them to oviposit, were examined in a three-year study of the hover flies and aphids occurring in a patch of broom bushes, with additional observations on some Scots pine trees (Dixon, 1959). The hover flies included *Syrphus luniger*, *S. vitripennis*, *S. nitidicollis*, *S.balteatus* (Plate V), *Platycheirus scutatus* (Meig.) and *Scaeva pyrastri* (L.) (Plate V). Two species of aphid occurred on the broom, both spending their entire lives on the bushes. The commonest was *Acyrthosiphon pisum* ssp. *spartii* (Koch), which was well-distributed amongst young shoots and leaves, the flowers and the pods, from about mid-April. The second species, *Aphis sarothamni* Franssen, was less numerous, forming a few dense colonies.

It was found that the flies tended to lay eggs on the bushes most heavily infested with aphids, laying their eggs very close to the aphids and laying more frequently where there were larger aphid colonies. The aphid populations became greatest during the flowering period of the broom, and the peak of abundance of hover fly eggs occurred before the peak in numbers of *A. pisum* ssp. *spartii*. Hover flies whose larvae feed on aphids are unusual among the Diptera in being able to take solid food, and a diet of pollen as well as nectar is required to ensure the ripening of the ovaries in the female. Since yellow objects are attractive to certain hover flies, it was suggested that the flies were attracted by the broom flowers because of their colour and the food supply they provided. Thus many eggs would be laid at a time when the aphid populations were rising, so that there was abundant food for the larvae. After the flowering period aphid numbers declined and fewer hover flies visited the bushes to lay eggs. Attraction to yellow does not mean that hover flies ignore flowers of other colours; they were, for example, seen visiting yarrow, vetch, creeping thistle and knapweed while the broom was in flower, and in searching for flowering plants would discover aphid colonies in a variety of situations. The likelihood of a female visiting an area probably depends upon the number, kind and colour of flowers present, as well as weather conditions.

The behaviour of several species during egg laying was also studied and there was a characteristic sequence of events. The female approached a stem, flying in a straight line from a distance of several yards. She then hovered up and down the stem until she found herself close to an aphid colony, over which she would hover for a short time before alighting with ovipositor extended. She would then walk over a small area of the plant, alternately extending and withdrawing her ovipositor and proboscis, and moving her antennae in a vertical plane; finally the ovipositor was bent downwards and drawn over the stem, and an egg was laid. Experiments to study the effects of various stimuli on *Syrphus luniger* suggested that a female ready to lay eggs was attracted by green colours, which would lead her to young plant growth where aphids are more likely to occur. The smell of the aphids was found to be the most important stimulus which caused the female to alight and lay eggs; this would facilitate the discovery of aphids in concealed positions, such as in curled leaves or galls—perhaps even on the roots of plants, for *Pipizella varipes* Meig. tends to lay eggs on wild parsnip plants whose roots are infested with aphids.

Neither adult nor young *A. pisum* ssp. *spartii* showed any apparent response to hover flies on the wing, but when a female alighted the adult aphids tended to point one or both antennae towards her, then walk away, while the first stage nymphs did not retreat or even withdraw their stylets. Consequently, hover fly eggs were more often laid close to colonies of young aphids, which was advantageous to the larvae when they hatched. Some eggs were found actually adhering to aphids. Except for *Melanostoma* species and *Platycheirus* species, which often lay eggs in batches of 2–5 at a time, the hover flies predatory on aphids tend to lay one egg beside each aphid colony they discover. Their progeny are probably thus well spread over aphid infestations.

In experiments to study the effects of ants guarding aphids against predators (Banks, 1962), it was found that the common black ant would protect black bean aphid against larvae of *Syrphus balteatus*, as it did against ladybird larvae. When attacked by ants the hover fly larvae protected themselves by a slimy exudation and, unlike the ladybird larvae, were not actually driven off the plants. The hover fly larvae also proved to be more effective predators than ladybirds, for in the absence of ants they ate aphids more quickly and reached small aphids hidden in the folds of young leaves where they usually escape attack by ladybird larvae. With one hover fly larva per aphid-infested bean plant, no aphids survived after six days, while with one larva of the two spot ladybird per plant only 20–40 per cent of the aphids were killed.

S. balteatus, *S. luniger* and *S. ribesii* have been recorded as predators of cabbage aphid (George, 1957). *Scaeva pyrastri* preys on the cabbage aphid and the pea aphid. The larvae of some species attack the fruit tree red spider mite (Collyer, 1953), and those of *Syrphus luniger* have been seen to tackle caterpillars of the lackey moth (*Malacosoma neustria* (L.)). Caterpillars of the small white butterfly (*Pieris rapae* (L.)) are also among the prey of some species.

In a survey of aphid predators on potatoes in northern England (Dunn, 1949), *Platycheirus manicatus*, *P. scutatus*, *P. immarginatus* (Zett.), *Sphaerophoria* species, *Syrphus balteatus* and *S. vitripennis* and several others were recorded. Hover fly eggs were generally found in numbers equivalent to two or three per 100 leaves, but sometimes more abundantly. *P. manicatus* and *P. scutatus* occurred regularly but did not exert much check on the aphid populations; *S. balteatus* was more infrequent but when it did appear it was more numerous and had the greatest control value of all the species. Eggs were found from mid-July to the end of August, so that the hover flies were not present until the aphid infestations were well advanced. Hover flies would nevertheless seem to be predators of considerable potential importance; more knowledge is needed of their effectiveness under different conditions and of factors affecting their abundance, including the decrease of wild flowers by the use of herbicides and the clearance of hedgerows and uncultivated land.

Other Predatory Flies

There are several other families of predatory flies. The Asilidae (robber or assassin flies) and Empididae with over 300 species between them are predatory as adults of both sexes and have mouthparts in the form of a long, sucking beak. The Asilidae are medium to large-sized, bristly flies with narrowly built but robust bodies, large compound eyes and long, bristly legs. Many of the species lie in wait on the ground or on vegetation, resting more or less on one side with their legs drawn up, reminiscent of a leopard crouching to spring. On the approach of a potential victim they make a sudden flying hop or 'capture dart'. The legs hang loosely during flight before the victim is seized, or in some species are extended forward like a scoop. The prey is often larger and stronger than the hunter and includes grasshoppers, beetles, bees, wasps, ichneumons and flies.

The Empididae are small or medium-sized, with narrow, bristly bodies, large compound eyes and long legs. They may hunt on the wing or by running rapidly over vegetation; their prey appears to consist mainly of other flies. *Tachydromia minuta* (Meig.) and *T. agilis* (Meig.) have been observed preying upon frit flies (Jones, 1965).

Larvae of the stiletto flies (Therevidae) are said to be predatory, living in the soil, leaf litter and rotten wood. These larvae have long snake-like bodies (Plate V), looking like white wireworms, and may be turned up by the gardener while digging.

The Phoridae is a large family, many of whose members live in fungi and decaying plant matter, but it also includes some predatory species. *Plastophora rufa* (Wood) has been found in Bavaria to lay its eggs among those of the brown scale (*Parthenolecanium corni* (Bouché)) on fruit and other trees, and the larvae which hatch from them feed on the eggs of the scale insect and then on the parent scales (Schmutterer, 1952).

Although the Chloropidae are mainly plant feeders in their larval stages, some being pests, such as frit fly and gout fly (*Chlorops pumilionis* (Bjerk.)), several species are known to prey upon aphids. Larvae of *Thaumatomyia glabra* (Meig.) and *T. notata* (Meig.) are important predators of lettuce root aphid and have been commonly found amongst colonies of this aphid on the roots of lettuce. They have two or three generations a year and each larva of *T. glabra* has been estimated to consume about 75 aphids in its lifetime (Dunn, 1960).

The family Sciomyzidae includes a number of species whose larvae prey upon snails, including aquatic species, and slugs. *Tetanocera elata* (F.) has been shown to feed exclusively on slugs (Knutson, Stephenson and Berg, 1965). Under laboratory conditions the females laid up to 373 eggs each. In the field the eggs are presumably laid in vegetation frequented by slugs, though the presence of the latter is not necessary to induce oviposition. After hatching, the young larvae do not crawl about as if in search of prey but wait by their empty eggs. They will die of starvation if no suitable victim appears in four or five days, but on making contact with a passing slug—and apparently only the field slug and the marsh slug (*Agriolimax laevis* (Müll.)) are acceptable—a larva quickly crawls on to it and, after an initial bite near the posterior end of the mantle, crawls down the side of the slug and penetrates its body beneath the edge of the mantle.

Infested slugs may survive a week or two depending upon the number of larvae they are supporting. Most infested slugs have been found to harbour from one to five larvae, occasionally as many as ten. In the first stage of its life the larva feeds on mucus but later, especially in the third stage, the larva becomes more predacious and leaving its dead host actively pursues further victims. These are usually seized at the head end and quickly immobilized through laceration of their nervous tissues. As soon at it is satiated the larva leaves its victim; a succession of four to nine slugs may be quickly attacked and fed upon by one larva in this third stage. At this time they are less specific in choice of food and will prey upon several other common species of slug. When fully fed the larva pupates in the soil, the total life cycle occupying two to three months. Two or three generations may occur during the year. A random sample of the field slug from a derelict horticultural plot contained 14 per cent which harboured larvae of this fly.

The family Cecidomyiidae, the gall midges, also includes some species with predatory larvae (Barnes, 1946–56). For example, *Arthrocnodax wissmanni* Kieffer preys upon the fruit tree red spider mite and gall mites. Larvae of *Feltiella tetranychi* Rübs. attack the red spider mite and have been found in numbers on hop leaves. A number of *Lestodiplosis* species are predatory, their larvae frequently preying upon those of other gall midges.

Larvae of *Aphidoletes* (=*Phaenobremia*) species have been recorded as preying upon cabbage aphid (George, 1957). The red, elongated eggs were found laid in groups of up to 40 beneath colonies of aphids, and each larva ate from 40 to 60 aphids. The insects were found quite commonly in two successive years in one Brussels sprout crop but not in another at a different site.

Tachinid Flies

Most of the parasitic flies belong to the family Tachinidae (Larvaevoridae). These are bristly flies in the suborder of the Diptera which includes the house fly, blowflies (bluebottles), greenbottles and flesh flies. More than 250 species are found in this country and the larvae of the majority are internal parasites of earthworms, snails, adult beetles, Orthoptera (grasshoppers, cockroaches, etc.), Hemiptera, Hymenoptera and flies, but most often of the caterpillars of butterflies or moths. One or two species attack the predacious centipedes, *Lithobius* species, and a small group parasitizes woodlice. A few are external parasites of mammals, for example the greenbottles or sheep blowflies (*Lucilia* species), and of nestling birds. The larvae, like those of other flies, are legless and their mouthparts are much reduced; the various ways in which they are enabled to secure themselves safely in the bodies of their hosts are of great interest.

Development of a piercing ovipositor as is characteristic of the Hymenoptera (page 28) is only found in a few species. *Compsilura concinnata* (Meig.), a common parasitic fly (Plate V), has a piercing organ with which it injects its larvae into the bodies of caterpillars. The simplest provision perhaps is the development in the body of the female fly of a means of incubating her eggs and larvae until they are ready to be deposited; species of this type are common. The larva is spiny to help it move over the body of the victim on which it is placed; movement is aided also by its mouth-hook with which it saws a point of entry. In flies such as *Dexia* species the larvae have an armoured skin and are extremely active; although so small, they can live some time without food and work their way down into the soil, through a distance of as much as a foot, in search of the larvae they are to parasitize. The flies appear to have some means of detecting situations containing suitable hosts. In another group of species, including some whose larvae parasitize woodlice, eggs are laid and from them hatch larvae active and armoured like those of *Dexia* species, capable of searching for their hosts. The latter, however, live in colonies under bark and such situations.

Species of *Tachina* and *Carcelia*, which deposit eggs instead of young larvae, provide the eggs with hard protective shells and adhesive properties. In *Tachina* they are convex with a flat base which is glued to the skin of the host caterpillar; the larva simply cuts through this base and bores straight into the body of the caterpillar. In *Carcelia* the eggs are cylindrical with a short stalk ending in a sticky expansion for adhesion to the host. The larva leaves the egg and finds its way into the host's body.

Another singular group, including the largest and most striking of tachinid flies, deposits on leaves larvae with a protective armour of plates and scales on their backs and sides. These larvae do not appear to search for a host but wait until a suitable caterpillar approaches, when they raise and wave their head end from side to side until they touch the caterpillar, which is seized and rapidly entered. Still more curious are species which lay their eggs on plants frequented by the host insect on the chance that they will be swallowed by it. The minute eggs are hemispherical with a thick pigmented upper surface and do not hatch until within the mid-intestine of a host caterpillar, from which they bore into the body cavity. The eggs can survive quite a long waiting period, but many must go uneaten or be crushed. As with the previous group the mortality rate of young stages must be high; the parasite compensates for this by producing enormous numbers of progeny. A female of an *Echinomyia* species dissected by the great French naturalist Réaumur was said by him to have contained no less than 20,000 larvae.

A few species of tachinid larvae live in the body cavity, apparently obtaining their oxygen through the skin like most parasitic hymenopterous larvae, while some begin their parasitic existence inside an organ such as a muscle fibre. Generally, however, there is provision for direct access to the atmosphere for breathing. The larvae breathe through a pair of openings, called spiracles, at the posterior end of the body. Some species, which bore into the body of the host from outside, leave their rear end in the entrance hole, where the host's skin, irritated by the injury, grows in round it forming a sheath which actually serves to secure the parasite's hold. Other species live for a time without connection with the exterior but later push an opening through the host's body wall. Others utilize the host's own air supply, tapping its system of respiratory tubes which obtain air through the spiracles and transfer it to the various parts of the body. Tachinid larvae are on the whole larger than those of hymenopterous parasites and it is unusual to find more than one or two individuals in one host. There are, however, a few species attacking caterpillars where large numbers of parasitic larvae occur together; as many as 90 have been found in one of the large bombycid caterpillars.

When fully fed, tachinid larvae either pupate inside the skin or cocoon of the host or they leave the host and enter the ground. As with many other flies there is a puparium, which is a round-ended, barrel-shaped object showing lines of segmentation but no indication of legs or wings such as in moth or beetle pupae. It is usually brownish or reddish-brown and is the skin of the larva, hardened and contracted to form a protective case inside which the actual pupa is formed. When the fly eventually emerges, the front of the head has a fold of skin which is armed with backward-pointing spines and which can be distended like a bladder by blood pressure. By alternately distending and contracting this organ the insect can draw itself up through the soil; having reached the surface it expands its wings, and the retracted bladder is enclosed by the hardening skin and never used again.

Tachinids hibernate in all stages except the egg, often within the body of the host insect. No dipterous parasites of wholly aquatic insects are known, though there are a few inhabiting the larvae of certain crane flies; these live in water but come to the surface periodically to obtain air.

Other Parasitic Flies

There are several other families of flies which include parasitic species. The Cecidomyiidae has several species whose larvae are internal parasites of aphids and suckers (Barnes, 1954). The Pipunculidae, a family of small dark flies with abnormally large heads and the habit of hovering—they are related to the hover flies—deposit their eggs in the bodies of leafhoppers and tree hoppers, within which the larva feeds and grows to fill the greater part of the abdomen. The family Phoridae also includes some parasitic individuals, though little is known about them.

References and Further Reading

Banks, C. J. (1953). The natural enemies of aphids. *Rep. Rothamsted exp. Stn for 1952*, 116–7.

Banks, C. J. (1962). Effects of the ant *Lasius niger* (L.) on insects preying on small populations of *Aphis fabae* Scop. on bean plants. *Ann. appl. Biol.* **50,** 669–79.

Barnes, H. F. (1946–56). *Gall midges of economic importance.* Vols. I–VII. London, Crosby Lockwood.

Barnes, H. F. (1954). Gall-midge larvae as endoparasites, including the description of a species parasitising aphids in Trinidad, B.W.I. *Bull. ent. Res.* **45,** 769–75.

Collyer, Elsie (1953). Biology of some predatory insects and mites associated with the fruit tree red spider mite (*Metatetranychus ulmi* (Koch)) in south-eastern England. III. Further predators of the mite. *J. hort. Sci.* **28,** 98–113.

Collyer, Elsie (1956). Some predacious insects of fruit trees: lacewings, ladybirds and hover flies. *Rep. E. Malling Res. Stn for 1955*, 155–6.

Colyer, C. N. and Hammond, C. O. (1968). *Flies of the British Isles.* The Wayside and Woodland Series. London and New York, Frederick Warne.

Dixon, T. J. (1959). Studies on oviposition behaviour of Syrphidae (Diptera). *Trans. R. ent. Soc. Lond.* **111,** 57–80.

DUNN, J. A. (1949). The parasites and predators of potato aphids. *Bull. ent. Res.* **40,** 97–122.

DUNN, J. A. (1960). The natural enemies of the lettuce root aphid, *Pemphigus bursarius* (L.). *Bull. ent. Res.* **51,** 271–8.

GEORGE, K. S. (1957). Preliminary investigations on the biology and ecology of the parasites and predators of *Brevicoryne brassicae* (L.). *Bull. ent. Res.* **48,** 619–29.

JONES, M. G. (1965). The effects of some insecticides on populations of frit fly (*Oscinella frit*) and its enemies. *J. appl. Ecol.* **2,** 391–401.

KNUTSON, L. V., STEPHENSON, J. W. and BERG, C. O. (1965). Biology of a slug-killing fly, *Tetanocera elata* (Diptera: Sciomyzidae). *Proc. malac. Soc. Lond.* **36,** 213–20.

SCHMUTTERER, H. (1952). *Plastophora rufa* (Wood) (Dipt., Phoridae) als Eiräuber und Parasit von *Eulecanium corni* (Bché) (Homopt., Coccoïdea). *Anz. Schädlingsk.* **25,** 145–8.

SUNDBY, R. A. (1966). A comparative study of the efficiency of three predatory insects *Coccinella septempunctata* L. (Coleoptera, Coccinellidae), *Chrysopa carnea* ST. (Neuroptera, Chrysopidae) and *Syrphus ribesii* L. (Diptera, Syrphidae) at two different temperatures. *Entomophaga* **11,** 395–404.

MITES

Although mites (Acari) are not insects, they must be given a place in this Bulletin as, like the insects, many have adopted predatory and parasitic ways of life and they form important elements in the natural control of some pests, especially the tarsonemid and spider mites. They differ in structure from insects in a number of ways; for example, they lack the distinct division of the body into the three regions, head, thorax, and abdomen, characteristic of adult insects, and the adults possess four instead of three pairs of jointed legs. They are small, inconspicuous creatures and the number of species is extremely large. Only a few are important pests of crops, for example fruit tree red spider mite, red spider mite, black currant gall mite (*Cecidophyopsis ribis* (Westw.)), strawberry mite (*Steneotarsonemus pallidus* (Banks)), bulb scale mite (*S. laticeps* (Halb.)), with several other species occurring as pests of mushrooms and stored foodstuffs.

The family Phytoseiidae includes some important predators; species of *Typhlodromus* have a potential value as agents in the biological control of spider mites on fruit trees, and recently considerable attention has been given to the South American species *Phytoseiulus persimilis* Athias-Henriot (= *P. riegeli* Dosse) for use in control of red spider mites on glasshouse crops (see page 94).

PREDATORY MITES ON FRUIT TREES

The phytoseiid mites occurring in Britain are mostly small, colourless or pale brown, and individuals of one kind or another may be found on almost every kind of plant. Their occurrence and biology on fruit trees has been studied at East Malling Research Station (Collyer, 1956; Chant, 1957; and Niemczyk, 1965). Nine species were found on unsprayed fruit trees (Collyer, 1956). Of these, *Typhlodromus pyri* Scheut. (= *T. tiliae* Oudms.) appeared to be the most important as it was the species occurring most abundantly on sprayed apple trees. It was found on apple, pear, plum, hop, raspberry, a wide variety of woodland and hedgerow trees and shrubs, and also herbaceous plants. The mites hibernate as adult females in crevices and under bark, especially on the trunk, some ten per cent of them surviving the winter. When activity is resumed in the spring, the mites move among the developing buds, showing a preference for sheltered places such as under loose bud scales, in crevices in the bursting buds, and later within and between the developing blossom buds. As the leaves expand, some of the mites move out on to their lower surfaces, and from petal fall onwards most are to be found on the foliage.

The ovoid, colourless eggs are laid singly on the underside of the leaves from late April onwards, often near the mid-rib and main veins. With *T. finlandicus* (Oudms.), the second most common species, the eggs are often laid at the tips of leaf hairs, and with some other species they may be laid in the silk webbing spun by red spider mites. The eggs of *T. pyri* hatch in about 4–8 days and the young mites pass through one larval and two nymphal stages before becoming adult in 9–14 days. The larva has only three pairs of legs but the nymphs, like the adults, have four pairs. Under laboratory conditions the adult females usually lived three or four weeks, sometimes nearly two months, and laid about 30 eggs. Usually three, sometimes four, generations occurred in the year; in the field there are probably three.

All stages of these predacious mites, except the larvae, attack and feed upon mites such as spider mites, gall mites and others. All active stages and summer eggs of the spider mites are eaten, except the eggs and older mites of *Bryobia* species, but the summer eggs of fruit tree red spider mite are found difficult to penetrate. Winter eggs have not been seen to be attacked. When feeding on very small mites, such as the apple and plum rust mites (*Aculus* species), one individual may consume up to 90 a day, but with a diet of fruit tree red spider mite the daily average is two or three.

Feeding on fruit tree red spider mite causes the predator's intestines to turn red and, using this as a criterion, it was found in one study (Chant, 1957) that the percentage of typhlodromid mites which had eaten the spider mite varied from about 20–70 and might on occasion be higher. The effect of alternative prey on the relationship between *Typhlodromus* species and fruit tree red spider mite has also been studied (Collyer, 1964) in experiments with cultures of mites on plum seedlings. It was found that *T. pyri* and *T. finlandicus* increased more rapidly when provided with the eriophyid mite *Aculus fockeui* (Nal.) as well as fruit tree red spider mite, and reduced populations of the latter to a lower level than it did in the absence of the alternative prey.

The mites can survive for long periods by feeding on the leaves, but an additional source of protein is needed for the young stages to complete their development (Chant, 1957). Animal protein is not necessary and a diet of pollen grains and fungal spores can satisfy their requirements.

The variety of food taken by these mites means that they are not dependent on a particular prey, so they do not necessarily become more numerous when their prey increases. On the other hand there is the advantage that they will remain on the tree, in contrast to insect predators which tend to move off elsewhere when they have decreased their prey to low numbers. Attempts to measure their effectiveness as predators of fruit tree red spider mite, which show that they may have a useful part to play, are discussed on page 85.

Anystis agilis Banks (family Anystidae) is also recorded as a predator of the fruit tree red spider mite (Collyer, 1953). This is a large red mite which moves rapidly over the leaves of fruit and other trees and bushes, attacking various mites and insects. It was found occasionally in sprayed orchards but more commonly on unsprayed trees. A familiar mite on account of its red colour and large size is the red velvet mite, *Allothrombium fuliginosum* (Herm.) (family Trombidiidae), often seen walking about the ground and in vegetation in the spring and summer. It feeds on small insects, including the woolly aphid on apple trees. It hibernates as an adult and lays its eggs in the soil in spring (Massee, 1954). Species of *Pergamasus*, *Eugamasus* and *Veigaia* prey upon springtails and mites in damp organic matter, and *P. longicornis* Berlese also eats the eggs of frit fly (Jones, 1967).

References

CHANT, D. A. (1957). Predacious spiders and mites on fruit trees. *Rep. E. Malling Res. Stn for 1956*, 175–8.

COLLYER, ELSIE (1953). Biology of some predatory insects and mites associated with the fruit tree red spider mite (*Metatetranychus ulmi* (Koch)) in south-eastern England. IV. The predator-mite relationship. *J. hort. Sci.* **28**, 246–59.

COLLYER, ELSIE (1956). Notes on the biology of some predacious mites on fruit trees in south-eastern England. *Bull. ent. Res.* **47**, 205–14.

COLLYER, ELSIE (1964). The effect of an alternative food supply on the relationship between two *Typhlodromus* species and *Panonychus ulmi* (Koch) (Acarina). *Entomologia exp. appl.* **7**, 120–4.

JONES, MARGARET G. (1967). Predators of frit fly. In: Johnson, C. G. Entomology department. *Rep. Rothamsted exp. Stn for 1966*, 203–4.

MASSEE, A. M. (1954). *The pests of fruits and hops*. London, Crosby Lockwood.

NIEMCZYK, E. (1965). The distribution of *Typhlodromus pyri* Scheut. (Acarina: Phytoseiidae) on apple trees in spring. *Rep. E. Malling Res. Stn for 1964*, 180–3.

Insect Diseases

ALTHOUGH this Bulletin is principally concerned with insects as control agents, it should be noted that there are numerous diseases of insects caused by micro-organisms of various kinds, e.g., bacteria, viruses, Rickettsia, Protozoa, and also fungi and eelworms. As with predatory and parasitic insects, problems associated with the use of insecticides have given a great stimulus in recent years to research on insect diseases, with considerable promise that some of them can be used as alternatives to chemicals or integrated with various other pest control measures.

BACTERIAL DISEASES

Of the bacterial diseases there is a group known as the milky diseases caused by several species of *Bacillus* and recorded from a variety of insects, including the Japanese beetle (*Popillia japonica* Newm.) and other chafer beetles. When infected with *B. popilliae* Dutky, the blood of larvae of the Japanese beetle becomes milky instead of almost clear as in healthy individuals, and the body becomes turbid and opaque. Diseased larvae usually die before pupation. The larvae may also be infected with *B. lentimorbus* Dutky, which causes the body to become a muddy brown colour instead of creamy white. Spore dusts of these bacteria have been successfully used in the U.S.A. for control of Japanese beetle in turf.

In another group of bacteria, a protein crystal is formed when the bacterium sporulates, and these proteins are highly toxic to some insects, especially larvae of many butterflies and moths (including the silkworm, *Bombyx mori* (L.)) but also larvae of some flies, beetles and Hymenoptera. One of the best known is *Bacillus thuringiensis* (Berliner) Mattes, and preparations containing this organism are now produced as dusts or sprays for application to crops. They have the advantage of being a selective stomach poison which does not affect predators or parasites; also they are not phytotoxic and do not leave residues poisonous to the human consumer. More experimental work is required before their effectiveness against pests in Britain can be established.

FUNGAL DISEASES

Many kinds of fungi are parasitic on insects. The family Entomophthoraceae includes the genera *Empusa*, *Entomophthora* and *Massospora* which are chiefly parasites of insects. *Entomophthora muscae* Cohn attacks the house fly, blowflies and hover flies. Affected flies may sometimes be seen adhering to window panes and surrounded by a halo of discharged fungus spores. *E. aphidis* Hoffm. and *E. exitalis* Hall & Dunn parasitize aphids. Colonies of the black bean aphid on beet and mangels are frequently destroyed by *Entomophthora* species in July, when under warm humid conditions the fungus spreads rapidly and heavy infestations are killed in a few days. Affected aphids turn from black to brown or cinnamon colour. There are also many species of *Cordyceps* which parasitize bugs, flies, moths, beetles and Hymenoptera. A stalk-like or club-shaped growth can be seen arising from the body of the victim. The 'muscardine' diseases are caused by species of *Beauveria*, affected insects becoming enveloped in a white or greenish fungal mat. *B. bassiana* (Bals.) Vuill. attacks larvae of codling moth (*Laspeyresia pomonella* (L.)) and other Lepidoptera including the silkworm, and other insects. There are several other groups of entomogenous fungi; 54 species known in Britain were listed by Petch (1932). Difficulties are likely to arise in attempts to use some of these fungi to control pests, as suitable temperature and humidity conditions, and a high density of the pest population, are often needed for them to become effective. When these favourable conditions do occur, a disease epidemic may arise naturally.

VIRUS DISEASES

Diseases of insects caused by various groups of viruses are known, especially in the larvae of butterflies and moths, but also in some flies, beetles and Hymenoptera.

Outbreaks of the European spruce sawfly (*Diprion* (= *Gilpinia*) *hercyniae*

(Hart.)) in Canada were reported as being largely controlled by a 'polyhedral' virus disease (Baird, 1948). There are a number of these polyhedral virus diseases—commonly called 'wilt diseases'—known in caterpillars and pupae of some Lepidoptera and Hymenoptera. In the U.S.A. it has been shown (Steinhaus, 1948; Steinhaus and Thompson, 1949) that the alfalfa caterpillar (*Colias eurytheme* Boisd.) can be controlled by spraying alfalfa (lucerne) with a suspension of infective virus material. Diseased caterpillars have a wilted or melted appearance after death, frequently remaining hanging by their prolegs (abdominal feet) from the plant. The disease spreads among caterpillars through contamination of the food plant by the disintegrating bodies of diseased individuals and by infected excrement; soil and surface debris can remain infective throughout the season. In a natural epidemic the disease kills large numbers of caterpillars, but frequently not until serious damage has already been done to the crop. By spraying crops with a preparation made from diseased caterpillars it has been found possible to initiate an epidemic resulting in almost 100 per cent mortality, though careful timing is necessary to ensure that control is obtained before economic crop damage can occur.

The discovery of a virus of the granulosis type infecting the caterpillars of the large white butterfly was reported by Smith and Rivers (1955). A caterpillar affected by the disease assumes a greasy or oily appearance and then hangs from the plant supported only by the first two pairs of prolegs, so that its body takes the form of an inverted V. The skin bursts and the contents of the body are scattered about its surroundings, thus rendering the food plant highly infective to other caterpillars. Spread of the virus is aided by the curious habit of the caterpillars of eating the bodies of their diseased fellows.

The virus can persist for at least 14 months in moist soil and for at least four months on cabbage leaves (David, 1965). It can be spread mechanically by the adult butterflies, flies, and the ovipositors of braconid parasites, and it can be transmitted from the parent butterfly through its eggs, although a caterpillar surviving infection does not necessarily produce an infective adult. The virus, which also affects the small white butterfly and the green-veined white butterfly (*Pieris napi* (L.)), has been reported as providing effective checks on infestations of caterpillars of white butterflies in various countries; it was suggested that an epidemic in Britain in 1955 resulted from an immigration of infected butterflies. Since virus preparations can be mass-produced for application as sprays to field crops, this virus offers attractive possibilities as a control measure. Not only is the virus highly specific so that predators and parasites would be spared, but there are no undesirable side-effects or toxic residues on the crop. A possible disadvantage is that the virus acts slowly, young larvae not dying within four days of infection and older ones taking longer, though they stop feeding some time before death. The persistence of the virus, however, might result in control of later generations. Also the spreading of infections over a wide area might, if ways of ensuring virus persistence through the overwintering insects into the following season could be found, result in virtual freedom from the pest for several years, as happened after the natural epidemic in 1955. Development of white butterfly strains resistant to the virus is a possibility, as this has occurred with laboratory cultures of the large white butterfly.

PARASITIC EELWORMS

Several species of eelworm parasitic in the bodies of insects are known, and more probably await discovery. An interesting example is that of *Bradynema megaselis* Welch occurring in a phorid fly (*Megaselia halterata* (Wood)) which is one of the serious pests of cultivated mushrooms. The eelworms live in the body of the adult, greatly reducing the capacity of the female to lay eggs, and in some populations up to 100 per cent of the flies are infected (Hussey and Gurney, 1963).

References and Further Reading

BAIRD, A. B. (1948). Biological control. *Rep. 5th Commonw. ent. Conf., 1948*, 60–3.

DAVID, W. A. L. (1965). The granulosis virus of *Pieris brassicae* L. in relation to natural limitation and biological control. *Ann. appl. Biol.* **56,** 331–4.

DEBACH, PAUL (Ed.) (1964). *Biological control of insect pests and weeds*. London, Chapman and Hall.

HUSSEY, N. W. and GURNEY, BARBARA (1963). Mushroom phorids. *Rep. Glasshouse Crops Res. Inst. for 1962*, 73–4.

PETCH, T. (1932). A list of the entomogenous fungi of Great Britain. *Trans. Br. mycol. Soc.* **17,** 170–8.

SMITH, K. M. and RIVERS, C. (1955). Virus attacks the cabbage caterpillars. *Grower* **44,** 932–5.

STEINHAUS, E. A. (1948). Polyhedrosis, ("Wilt disease") of the alfalfa caterpillar. *J. econ. Ent.* **41,** 859–65.

STEINHAUS, E. A. and THOMPSON, C. G. (1949). Preliminary field tests using a polyhedrosis virus to control the alfalfa caterpillar. *J. econ. Ent.* **42,** 301–5.

Pollinating Insects

SOME plant species can be fertilized with pollen from their own flowers, but often the flowers are so constructed that an external agent such as an insect or wind is needed to self-pollinate them, that is to transfer pollen from their anthers to their stigmas. Flowers of other species need to receive pollen from other plants (to be cross-pollinated) before they set seed or reach their maximum seed-producing capacity, so again an external agent is needed to carry the pollen. Wind is the principal pollinating agent of grasses, coniferous trees and a few other species, whereas it is believed that most plants with conspicuous, coloured and scented flowers are adapted for insect pollination.

POLLINATION REQUIREMENTS OF CROPS

To find whether a particular crop benefits from insect pollination, parts of the crop are enclosed in insect-proof cages, and honey bee (*Apis mellifera* L.) colonies put in some cages but not others. Sometimes (e.g., with brown mustard and dwarf beans) there is little or no difference in the amount of seed produced in plots caged with and without bees, indicating that automatic pollination or possibly wind pollination is sufficient to ensure a good crop. Sometimes (e.g., with white mustard) the presence of bees increases the seed yield although a considerable set occurs without them. However, many crops (e.g., carrots, onions, red clover, runner beans) produce little or no seed in cages from which insects are excluded. Even when there is little difference in the yield of plots with and without bees, insect pollination may have other advantages. Thus, field bean and broad bean plants visited by bees set a greater proportion of their earlier flowers than plants not visited, resulting in an earlier and more uniform crop; moreover field bean plants from seed produced by cross-pollination are auto-fertile, whereas those grown from seed produced from self-pollination are not. Although bee visits to strawberry flowers only slightly increase the percentage of them producing berries, there is a decrease in the percentage of berries that are malformed.

However, with the exception of glasshouse crops, the results of comparing caged plots with and without bees are not necessarily relevant to the increase that might result from taking honey bee colonies to a crop, as there may be pollinating insects, including honey bees, already present, or the crop may be relatively unattractive to bees compared with other flower species available.

Most species and varieties of fruit tree need pollination and there is much evidence that increased crops are obtained by moving honey bee colonies to orchards previously having few pollinating insects. Whereas varieties of peach, nectarine, apricot and sour cherry set fruit with their own pollen, most apple, pear, plum and sweet cherry varieties need cross-pollination or produce more and better quality fruit when cross-pollinated. In orchards of a variety needing cross-pollination it is usual to interplant another variety (called the pollinizer or pollinator variety) to provide the necessary fertilizing pollen. Recent experiments have failed to produce evidence that wind plays any part in the pollination of fruit trees.

TYPES OF INSECT POLLINATORS

Many species of insect visit the flowers of our economically important crops but, apart from the bees, most are beetles or flies. However, the various beetles that visit flowers are nearly all injurious and any inadvertent pollination may be more than counteracted by the harm they do. The hover flies (Syrphidae) are probably the most abundant pollinators among the Diptera, although the blowflies (Calliphoridae) are also important (Plate VIII). Because blowflies can easily be reared in captivity, they have been used by plant breeding stations for many years to pollinate plants in cages and glasshouses when attempting to produce new hybrids and varieties of crops such as carrots and onions. Nowadays there is a readily available supply of blowfly maggots produced as bait for fishing, and bait producers are usually willing to supply pupae from which adult flies will emerge in a day or so. Many growers who produce their own brassica seed from selected plants grown in insect-proof cages find that blowflies are cheaper and less bother to use than honey bees; Faulkner (1962) and Wiering (1964) found that they were as efficient. Although blowflies are suitable for pollinating open flowers, they are probably of little use in pollinating flowers with long corolla tubes.

The bees, i.e., solitary bees, bumble bees and honey bees, are our most valuable and abundant flower pollinators (Plates VI and VII). Their hairy bodies, which are adapted for collecting pollen, and their tendency to keep to one species while foraging, especially favour them as pollinators.

The solitary bees differ from the social bumble bees and honey bees in that the individuals do not live in communities: each female works alone in building her nest and providing food for her young, although several females may nest in close proximity to each other. Some solitary bees (e.g., species of *Colletes*, *Andrena* and *Halictus*) make their nests in burrows in the ground while others (e.g., species of *Megachile* and *Osmia*) build them in hollow stems of bramble, elder or decayed wood, using clay or vegetation to close the cells (Plate VIII). Various species of solitary bee may be locally abundant and help in pollinating our crops, particularly those with fairly open flowers, e.g. tree fruits, strawberry, raspberry, loganberry and blackberry (Plate VII).

Bumble bees are in general more abundant and widely dispersed than solitary bees. Although they live in colonies, these die out in the autumn and only the queens survive the winter and are available to pollinate fruit flowers in the spring. Nevertheless their large body size, speed of working, and ability to forage during weather conditions in which honey bees and solitary bees stay at home, make them most valuable pollinators. Whereas most, if not all, bumble bee species are valuable pollinators of open flowers such as those of blackberry, raspberry, and fruit trees, only those bumble bee species with long tongues (e.g., *Bombus agrorum* (F.) and *B. hortorum* (L.)) are able to obtain nectar from flowers with long corolla tubes such as runner beans and red clover (Plate VII). Many attempts have been made to induce bumble bees to found their colonies in captivity so that these can be moved to crops needing pollination (Holm, 1966), but even if this is successful it is doubtful whether bumble bee-keeping would ever be economically worth while except for the pollination of crops in cages and glasshouses. When the enclosures are small, bumble bees are preferable to honey bees and have been used for pollinating red clover strains at the Welsh Plant Breeding Station at Aberystwyth for many years.

The number of wild pollinating insects fluctuates from year to year and from place to place, but it has been suggested that there has been a general decrease in the number of wild pollinating insects in Britain as a result of the application of insecticides, which kill the pollinators themselves, and herbicides, which kill the wild flowers on which they feed. There is no sound evidence that such a decrease has occurred. More intensive cultivation of the land and removal of hedgerows must have decreased the number of nesting sites available for wild pollinators, although whether or not lack of nesting sites limits their numbers remains to be shown.

Under natural conditions there is usually no great concentration of one flower species in any one place, and the native insect population that is visiting them is probably sufficient to pollinate them. When many acres are occupied by a single flowering crop and many crops exist in the same locality, there may be too few wild pollinators, so that however favourable the other factors involved in producing a normal seed or fruit crop may be, the yield may be limited by lack of pollination.

HONEY BEES AS POLLINATORS

A shortage of wild pollinators may be overcome by moving honey bee colonies to crops needing pollination; in recent years this practice has grown considerably. Many County Beekeeping Instructors put growers in touch with beekeepers who have colonies for hire, and the Honey Producers' Association runs a similar but nation-wide service. The value that a grower will derive from hiring honey bee colonies depends on many factors including the number of bees already present. The presence of colonies beside or in the crop is particularly advantageous during inclement weather, because this will ensure the presence of pollinators during the short time available for flight; Ribbands (1951) showed that in unsettled weather bees do not forage far from their hives, so the grower cannot rely upon bees from distant colonies. Furthermore, recent work has shown that soon after a fruit flower opens, its ovules become less receptive, so there must be enough insects to pollinate the flowers within a day or so of opening (Williams, 1966). However, the number of honey bee colonies kept in England and Wales has decreased greatly in recent years and it is important that they should be used as efficiently as possible.

During summer the honey bee colony consists of a fertile female (the mother or queen), a few hundred males (drones), and several thousand sterile females (workers), on a series of wax combs on which the young are reared and the food is stored. After a varying length of time spent on household duties, all worker bees become foragers and visit plants to collect nectar and pollen.

It has long been known that a honey bee forager nearly always keeps to one species of flower during one trip; this habit obviously increases the chances of the flowers being pollinated and benefits the bee by enabling it to work faster. A bee also tends to visit the same species for several trips: Free (1963) found that 70–90 per cent of individual foragers collected the same kind of pollen over two consecutive days and about 40–60 per cent did so for seven consecutive days.

A bee does not wander at random over a crop but tends to forage in only a small part of it during one foraging trip. Crane and Mather (1943) grew two varieties of radish in adjacent blocks and found there was 30–40 per cent intercrossing where the two varieties joined, but this rapidly decreased to about one per cent where varieties were separated by 15 ft or more; they concluded that varieties should be separated by about 300 ft to avoid contamination. Bateman (1947) obtained similar results with turnip and radish varieties.

While small foraging areas may be advantageous to the seed grower who wants to avoid contamination between varieties growing in adjacent blocks, they may be a disadvantage to the fruit grower with orchards of self-sterile varieties, in which it is necessary to plant a pollinizer variety to pollinate the main variety. When foraging in orchards of standard fruit trees, a bee visits an average of only about two trees per trip and when it changes trees it tends to move to an adjacent one. In orchards where a greater distance separates rows than trees within a row, the bees move much more readily between trees in the same row than between trees in different rows. Bees also favour the sunny sides of trees. All these factors can be reflected in fruit set (Free, 1962). The sides of trees facing pollinizer trees sometimes have a greater set than their far sides, indicating that when a bee moves from a pollinizer to a main variety tree it pollinates only the first few flowers it visits; presumably the compatible pollen it carries is soon removed from its body or greatly

PLATE I

(Drawn by B. D. Moreton)

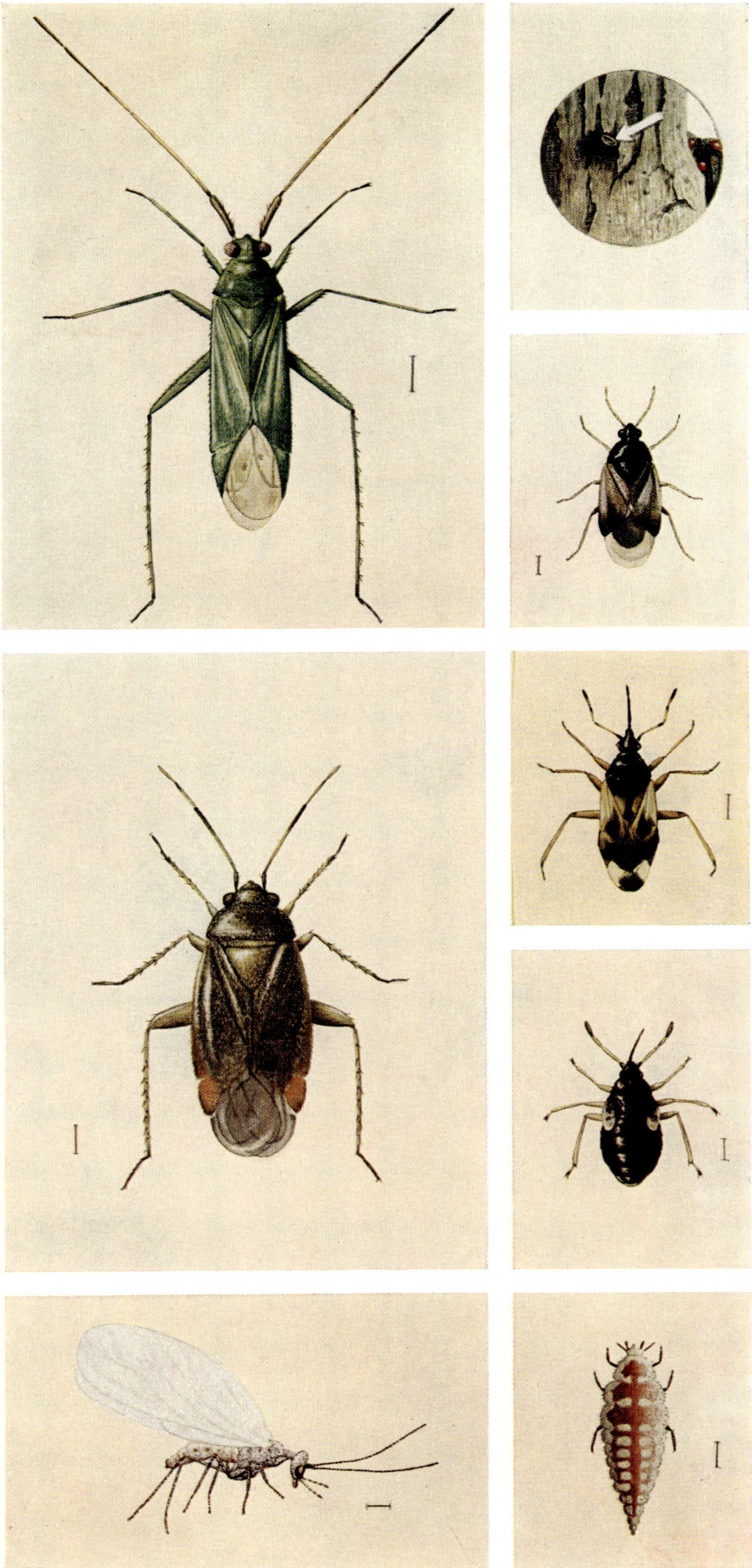

Left: Black-kneed capsid (*Blepharidopterus angulatus*)

Right (upper): Black-kneed capsid egg in plum shoot with cap showing (× 16)
Right (lower): Anthocorid bug, *Orius minutus*

Left: Red apple capsid (*Psallus ambiguus*)

Right (upper): Anthocorid bug, *Anthocoris nemorum*
Right (lower): Nymph of *Anthocoris nemorum*

Left: Powdery lacewing, *Conwentzia pineticola*

Right: Larva of a powdery lacewing, *C. pineticola*

PLATE II

Right: Green lacewing, *Chrysopa carnea* (Redrawn by B. D. Moreton)

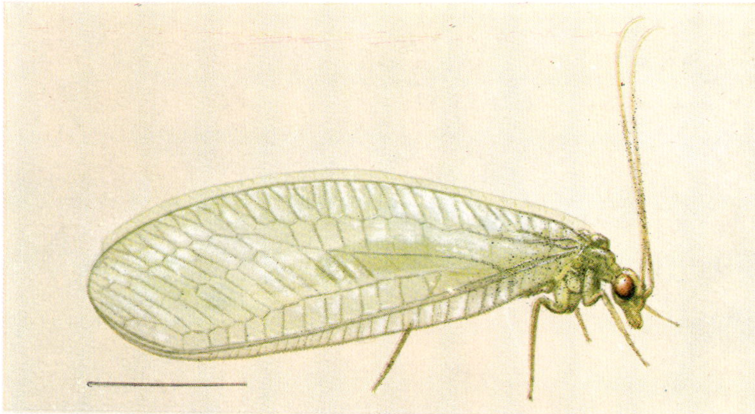

Left: Larva of a green lacewing, *C. carnea* (Redrawn by: B. D. Moreton)
Right: Cocoon of a green lacewing, *C. carnea* (Redrawn by B. D. Moreton)

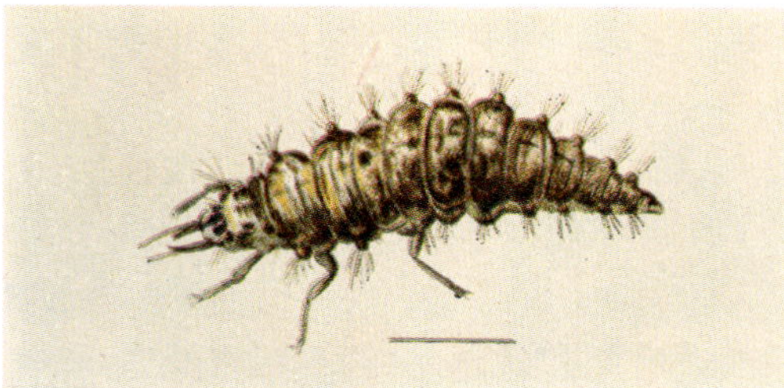

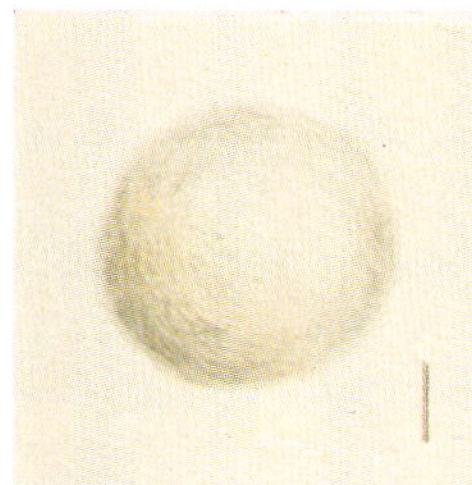

Left: Ground beetle, *Pterostichus melanarius* (Drawn by B. D. Moreton)
Right: Eggs of a green lacewing (×4) (Drawn by B. D. Moreton)

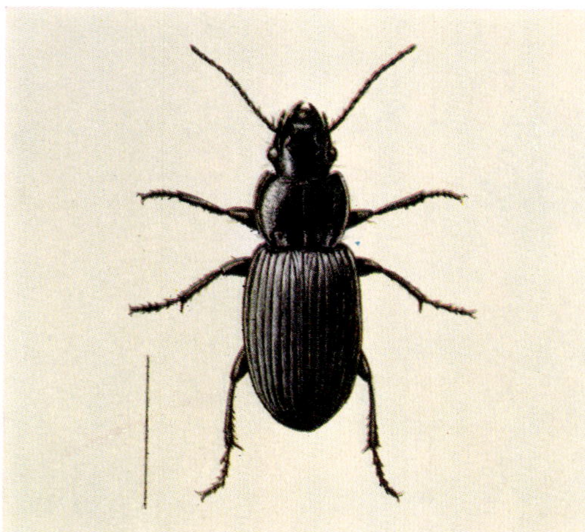

Left: Violet ground beetle (*Carabus violaceus*) (Drawn by B. D. Moreton)
Right: Larva of a ground beetle, *Carabus* sp. (Drawn by B. D. Moreton)

PLATE III

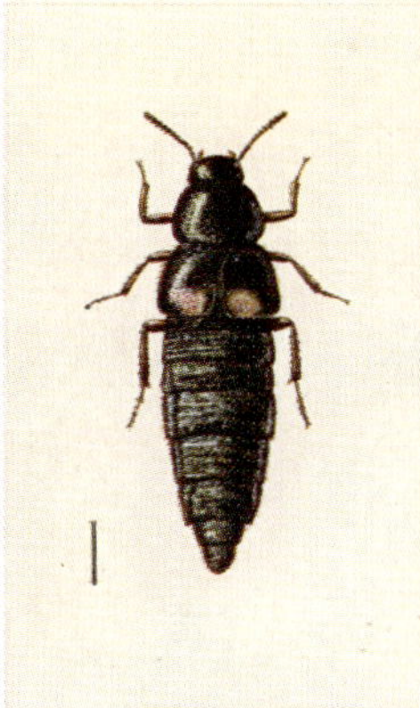

Left: Ground beetle, *Harpalus aeneus* (Drawn by B. D. Moreton)
Centre: Rove beetle, *Aleochara bipustulata* (Drawn by B. D. Moreton)
Right: Larva of a rove beetle, *Philonthus* sp. (Drawn by B. D. Moreton)

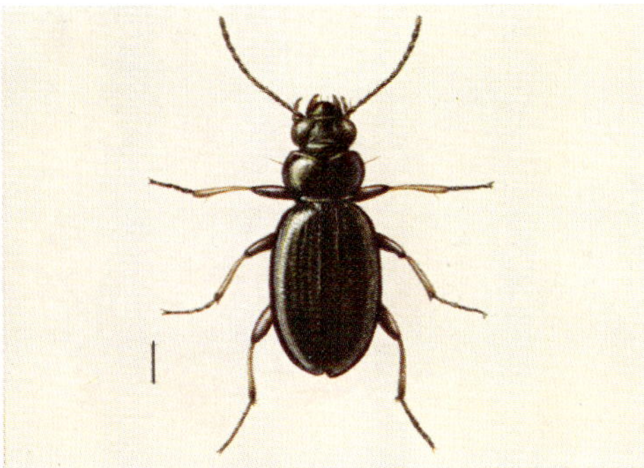

Left: 14 spot ladybird (*Propylea 14-punctata*) (Drawn by B. D. Moreton)
Right: Ground beetle, *Bembidion lampros* (Drawn by B. D. Moreton)

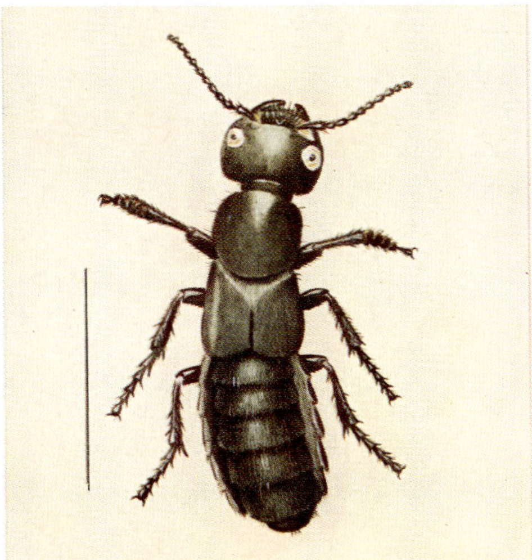

Left (upper): Ladybird, *Stethorus punctillum* (× 15) (Drawn by H. B. Foster)
Left (lower): 11 spot ladybird (*Coccinella 11-punctata*) (Drawn by H. B. Foster)
Right: Devil's coach-horse (*Staphylinus olens*) (Drawn by H. B. Foster)

Left: Two spot ladybird (*Adalia bipunctata*) (Drawn by H. B. Foster)
Centre: Larva of two spot ladybird (*A. bipunctata*) (Redrawn by B. D. Moreton)
Right (upper): Pupa of two spot ladybird (*A. bipunctata*) (Redrawn by B. D. Moreton)
Right (lower): Eggs of two spot ladybird (*A. bipunctata*) (× 4) (Redrawn by B. D. Moreton)

PLATE IV

Right: Red ichneumon, *Ophion luteus* (Redrawn by B. D. Moreton)

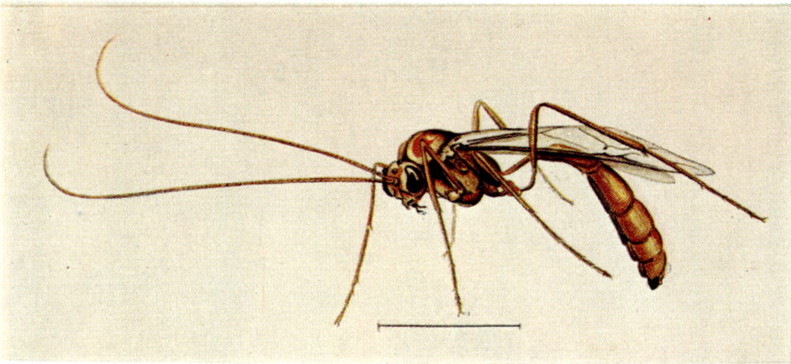

Left: Braconid wasp, *Apanteles glomeratus* (Redrawn by B. D. Moreton)
Right (upper): Cocoons of *Apanteles glomeratus* (× 2) (Redrawn by B. D. Moreton)
Right (lower): Larva of an ichneumon, *Paniscus* sp., attached by its stalked egg to the body of a caterpillar (× 4) (Drawn by B. D. Moreton)

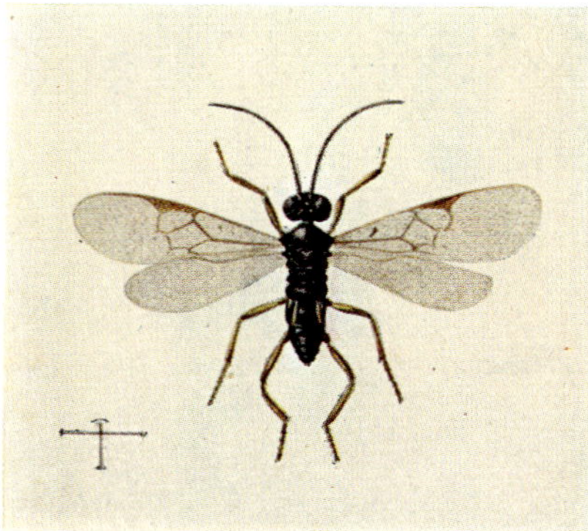

Right: Red-legged ichneumon (*Pimpla instigator*) (Redrawn by B. D. Moreton)

Left: Chalcid wasp, *Blastothrix sericea* (Redrawn by B. D. Moreton)
Right: Glasshouse whitefly parasite (*Encarsia formosa*) (× 25) (Drawn by B. D. Moreton)

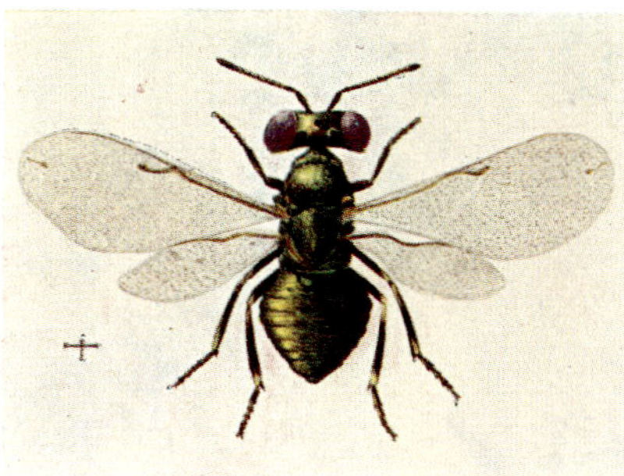

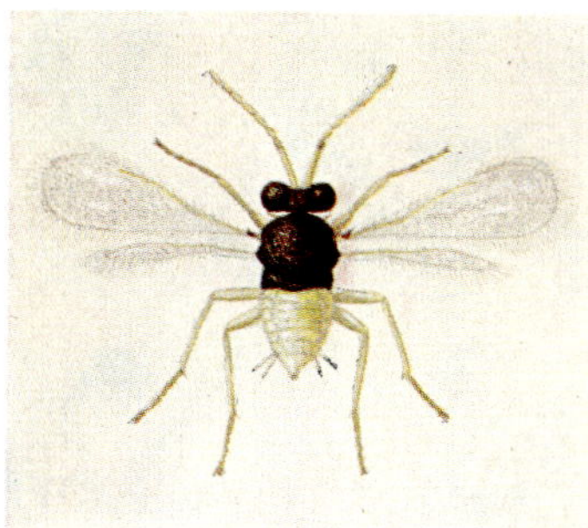

Left: Cabbage aphids parasitized by *Aphidius* sp. (× 8) (Drawn by B. D. Moreton)
Right: Glasshouse whitefly scales parasitized by *E. formosa* (× 15) (Drawn by B. D. Moreton)

PLATE V

Left: Hover fly, *Scaeva pyrastri* (Drawn by B. D. Moreton)

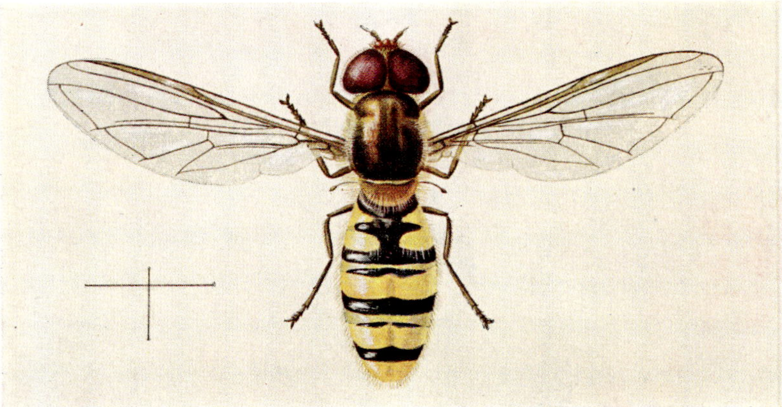

Left: Hover fly, *Syrphus balteatus* (Redrawn by B. D. Moreton)

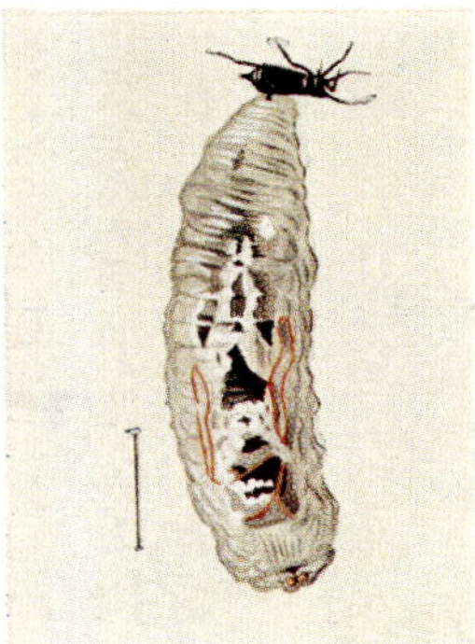

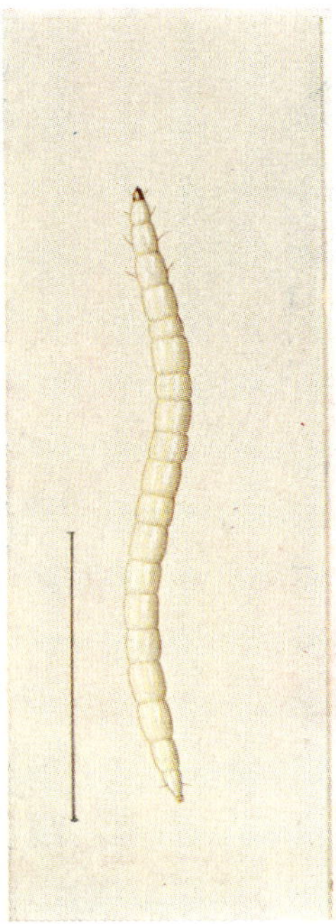

Left: Larva of *Syrphus balteatus* (Redrawn by B. D. Moreton)
Centre: Puparium of *Syrphus balteatus* (Redrawn by B. D. Moreton)
Right: Larva of a stiletto fly, *Thereva* sp. (Drawn by B. D. Moreton)

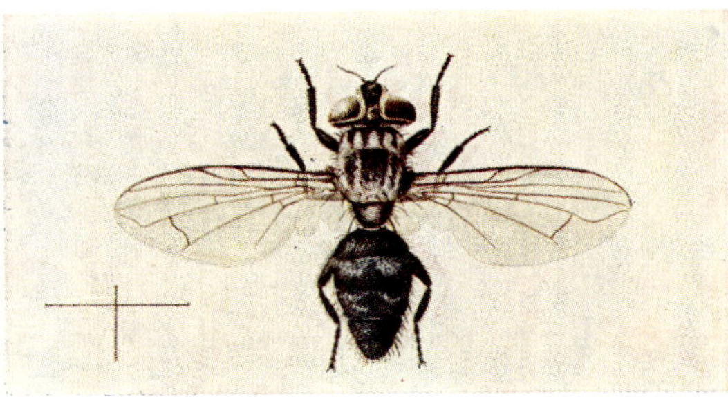

Left: Tachinid fly, *Compsilura concinnata* (Redrawn by B. D. Moreton)

PLATE VI
(Photos: J. B. Free)

Left: Honey bee on apple blossom (nectar gathering) (× 1)
Right: Honey bee on apple blossom (scrabbling for pollen) (× 1)

Left: Honey bee on bean flower (robbing nectar) (× 2)
Right: Honey bee on bean flower (pollen gathering) (× 2)

Left: Honey bee on rape flower (× 2)
Right: Honey bee on raspberry flower (× 2)

Right: Honey bee on strawberry flower (× $1\frac{1}{2}$)

PLATE VII
(Photos: J. B. Free)

Left: Bombus lucorum and nest ($\times \frac{1}{2}$)

Left: Bombus agrorum on blackberry flower ($\times$ 2)
Right: Bombus pratorum on raspberry flower ($\times$ 2)

Left: Bombus agrorum and nest ($\times$ 2)
Right: Bombus hortorum on clover flower ($\times 1\frac{1}{2}$)

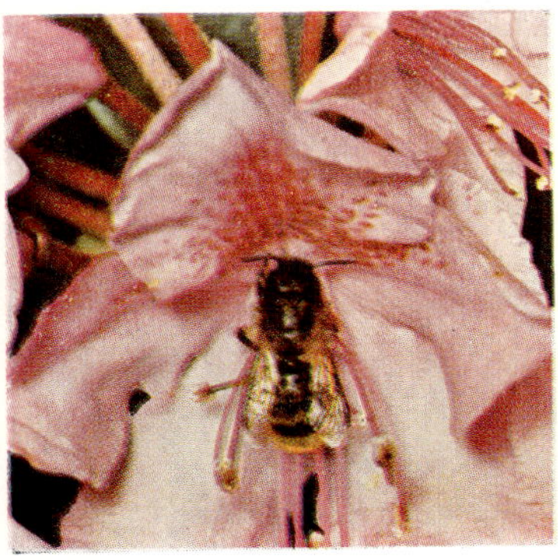

Left: Osmia rufa on rhododendron flower ($\times 1\frac{1}{2}$)
Right: Andrena haemorrhoa (male) on plum blossom ($\times$ 1)

PLATE VIII

Left: Blowfly on Brussels sprout flower (× 2)
Right: Hover fly (*Eristalis* sp.) on Strawberry flower (× 1)

(Photos: J. B. Free)

Left: Hover fly (*Scaeva* sp.) on pear blossom (× 1½)

Right: *Osmia rufa* building nest (× 1)

(Photos: J. B. Free)

Right: *Phytoseiulus persimilis*—egg and young (× 14)

(Photo: J. R. Morrison)

diluted with main variety pollen. To ensure even and, in some circumstances, adequate pollination in an orchard, each main variety tree needing cross-pollination should ideally be surrounded by pollinizer trees.

Although bees keep to small foraging areas on any one trip, during consecutive trips they may range quite widely in the orchard and visit different varieties. Bees do not always clean the pollen from their bodies thoroughly between trips, so pollen collected on one trip may be used to cross-pollinate on the next, and a bee may sometimes distribute pollen more widely than on a single trip (Free and Durrant, 1966). Pollen may also be transferred from bee to bee as they brush up against each other while in the hive, so a bee may cross-pollinate even when it remains constant to one variety for consecutive trips. This pollen transfer inside the hive could help explain such contamination as does occur in fruit and seed crops some distance from the pollinizer source.

When visiting open flowers like blackberry, strawberry and raspberry (Plate VI), both nectar- and pollen-gatherers touch the stamens and stigmas and so can transfer pollen to the stigmas. However, nectar-gatherers tend to keep their bodies more free of pollen than pollen-gatherers and hence are less likely to pollinate. All honey bees visiting flowers of fruit trees for pollen scrabble over the anthers and stigmas and so may pollinate, but whether or not nectar-gatherers pollinate fruit trees depends on where they stand on the flowers (Free, 1960) (Plate VI). Thus a nectar-gatherer that stands on the anthers and pushes its tongue and the front part of its body towards the nectaries touches the stamens and stigmas and may pollinate, but a nectar-gatherer that stands on the petals and pushes between the stamens to reach the nectaries does not touch the stigmas and so cannot pollinate, although when visiting flowers with spreading stamens (e.g., plum, pear, peach, apricot) it may get pollen on its body. When approaching from their sides flowers of apple varieties that have long stiff upright stamens, a nectar-gatherer does not touch the anthers and so does not even get pollen on its body. The proportion of nectar-gatherers approaching apple flowers from the side varies with the variety of apple and increases with the height and thickness of the stalks of their stamens, because the bees have difficulty reaching the nectaries from the top of the flowers.

Bees have difficulty in reaching the nectar in many flowers with long corolla tubes (e.g., red clover, field beans and runner beans) and some bumble bee species with relatively short tongues (*Bombus lapidarius* (L.), *B. lucorum* (L.), *B. terrestris* (L.)) bite holes in the bases of these flowers to reach the nectar; honey bees learn to use these holes as well (Plate VI) and so obtain nectar without pollinating, though they do not bite holes themselves. Some bees also visit the extrafloral nectaries of field beans and so again do not pollinate. Only those bees entering the flowers, most of which collect pollen, can pollinate them.

MANAGEMENT OF HONEY BEE COLONIES FOR POLLINATION

Bees tend to be constant to one particular manner in visiting a flower species, and because pollen-gatherers are usually more efficient than nectar-gatherers in pollinating the flowers, sometimes much more so, their numbers should be encouraged. This may be done most effectively by feeding colonies with sugar syrup (made by dissolving sugar in an equal weight of water) inside their hives (Free, 1965), although such feeding should be done by an experienced beekeeper only, as if done carelessly it could lead to robbing of each other's colonies of their honey stores. Although adult bees eat some pollen, most of that collected is needed to feed the developing brood and the amount collected is related to the amount of brood present. Colonies used for pollination should contain not less than five combs of developing brood of all stages.

When taken to a particular crop needing pollination, many bees visit plant species other than those intended, the proportion that does so varying from colony to colony. One way to increase the proportion of foragers visiting a crop is to delay taking the colony to it until flowering begins, so preventing bees becoming conditioned to other flower species which they do not readily forsake (Free, Free and Jay, 1960). Another way to increase the proportion of foragers visiting a crop is to eliminate competing flowers; thus dandelion is a serious competitor with fruit for bee visits and sometimes as many bees work the dandelions on the orchard floor as the trees themselves.

It is difficult to estimate the number of colonies needed to pollinate one acre of crop or the increase in yield that is likely from importing honey bee colonies, as these will depend on many factors including the density of flowers, amount of nectar and pollen produced, competition from other crops, weather and the number of pollinators already present. However, it is usually recommended that there should be one honey bee colony per acre of fruit orchard; recent investigations on apple have indicated that this is about correct.

Bees tend to forage near their hives so that the number of foragers on trees decreases quite rapidly as the distance of the trees from the colonies increases. Perhaps, ideally, colonies should be placed singly, each in the centre of its own acre, but it is more convenient to the grower and beekeeper to place colonies in groups. It has been found that when colonies are placed in groups of four in fruit orchards, at the rate of one colony per acre, the overlap in the foraging range of adjacent groups is sufficient to prevent any decrease in the number of foragers midway between them (Free and Spencer-Booth, 1963), but larger sized groups should be avoided.

References and Further Reading

Bateman, A. J. (1947). Contamination in seed crops. III. Relation with isolation distance. *Heredity* **1,** 303–36.

Crane, M. B. and Mather, K. (1943). The natural cross-pollination of crop plants with particular reference to the radish. *Ann. appl. Biol.* **30,** 301–8.

Faulkner, G. J. (1962). Blow-flies as pollinators of bassica crops. *Comml Grow.* No. 3457, 807–9.

Free, J. B. (1960). The behaviour of honeybees visiting flowers of fruit trees. *J. Anim. Ecol.* **29,** 385–95.

Free, J. B. (1962). The effect of distance from pollinizer varieties on the fruit set on trees in plum and apple orchards. *J. hort. Sci.* **37,** 262–71.

Free, J. B. (1963). The flower constancy of honeybees. *J. Anim. Ecol.* **32,** 119–31.

FREE, J. B. (1965). The effect on pollen collection of feeding honey-bee colonies with sugar syrup. *J. agric. Sci., Camb.* **64**, 167–8.

FREE, J. B. and DURRANT, A. J. (1966). The transport of pollen by honey-bees from one foraging trip to the next. *J. hort. Sci*, **41**, 87–9.

FREE, J. B., FREE, NANCY W. and JAY, S. C. (1960). The effect on foraging behavior of moving honey bee colonies to crops before or after flowering has begun. *J. econ. Ent.* **53**, 564–6.

FREE, J. B. and SPENCER-BOOTH, YVETTE (1963). The foraging areas of honey-bee colonies in fruit orchards. *J. hort. Sci.* **38**, 129–37.

GROUT, R. A. (Ed.) (1963). *The hive and the honey bee.* Hamilton, Illinois, Dadant.

HOLM, S. N. (1966). The utilization and management of bumble bees for red clover and alfalfa seed production. *A. Rev. Ent.* **11**, 155–82.

MEEUSE, B. J. D. (1961). *The story of pollination.* New York, Ronald Press Company.

PERCIVAL, MARY (1965). *Floral biology.* Oxford, Pergamon Press.

RIBBANDS, C. R. (1951). The flight range of the honey-bee. *J. Anim. Ecol.* **20**, 220–6.

WIERING, D. (1964). The use of insects for pollinating brassica crops in small isolation cages. *Euphytica* **13**, 24–8.

WILLIAMS, R. R. (1966). Pollination studies in fruit trees. III. The effective pollination period for some apple and pear varieties. *Rep. agric. hort. Res. Stn Univ. Bristol for 1965*, 136–8.

Pesticides and Beneficial Insects

PREDATORS AND PARASITES

AN account has so far been given of some of the better known predatory and parasitic insects and mites to be found in Britain. Most pests are probably subject to attack by a variety of natural enemies and the combined effect of these on certain pests where it is to some extent known will be described in a later section (page 75). Following sections will discuss the deliberate use of predators and parasites in securing pest control, and attempts to harmonize the use of chemicals with the activities of natural controlling agents. Before going thus far, however, it is necessary to consider the harmful effects of many pesticides on parasites and predators and their consequent influence on the balance between pests and their natural enemies.

Since the mid-1940s the use of chemicals for the control of insect and related pests, fungal diseases and weeds has increased in both variety and scale. Before that time there were relatively few insecticides available and except in horticulture their use was not very wide. Their lethal effects on insects and mites were limited and generally of short duration. By contrast, many modern insecticides are toxic to a wide range of insects and most persist for periods varying from a week or two to several months or even years. Although these new chemicals have brought enormous benefits by preventing or reducing losses by pests and diseases to crops in the field and in store, and by controlling the parasites of man and domestic animals and the spread of animal diseases carried by insect, mite and other vectors, their use has also brought new problems. In addition to the risks of contaminating foodstuffs with pesticide residues and incidental destruction of various forms of wild life apart from insects, problems have arisen with a number of pests which appear no easier to solve than those which faced us before the advent of these powerful new weapons. These problems may be described under three main headings.

(1). *Development of Resistance to Pesticides*. There are now numerous instances where the repeated use of a chemical against a pest has resulted in the selection of a strain of the pest resistant to the particular chemical used. No population of animals consists of individuals all exactly alike and, if a small proportion is tolerant of a pesticide, these individuals can survive and multiply until they are in the majority, thus forming a population against which the chemical is no longer effective. This is particularly likely to happen where a pest has several generations a year on one crop and a potentially high breeding rate. In the mid-1950s it was estimated that over 20 of the world's insect and related pests of economic importance had developed resistance to certain chemicals in some countries (Ripper, 1956). Examples from Britain are red spider mite, fruit tree red spider mite, cabbage root fly and peach-potato aphid on glasshouse chrysanthemums. These are discussed in more detail below. Resistance to pesticides is likely to come about more rapidly where the chemicals used also kill predators and parasites, so that natural checks on the surviving pest individuals are removed or reduced.

(2). *Development of New Pests*. Insects or mites of little or no economic importance may become numerous enough to constitute pests through destruction of their predators and parasites by chemicals directed against an existing pest. The well known example in this country is that of the fruit tree red spider mite, which will be referred to again later (see page 82).

(3). *Pest Resurgence*. Another effect of the destruction of predators and parasites is the resurgence or 'flare-back' of a pest, which may happen when the pesticide used kills the natural enemies but does not give a sufficiently high degree of control of the pest against which it is directed. This is especially likely with pests which have a succession of generations annually on the same host plant, such as spider mites, aphids and scale insects. In 1956 it was estimated that about 50 of the 5,000 or so major world pests had shown resurgence of this kind (Ripper, 1956). Examples from Britain will be discussed later.

It is not known whether the widespread use of pesticides has a further and more insidious effect by decreasing the number and variety of beneficial insects in areas of the countryside. Treatment of large acreages with insecticide may not only destroy predators and parasites in the crop but also form traps for insects moving in from adjoining land. Some aspects of this problem are being investigated by the Nature Conservancy and are referred to again on page 78.

The destruction of wild flowers by herbicides may also have adverse effects on useful insects. Van Emden (1965b), studying populations of cabbage aphid and its natural enemies in a field of Brussels sprouts with two sides bordered by trees, found that eggs of hover flies, the more important predators, were one and a half times more numerous on plants near the headlands with wild flowers than on plants in the centre of the field or on its open headlands where these wild flowers were absent. Ladybirds also appeared to have entered from the boundary vegetation, but the distribution of parasitic wasps seemed to be unaffected by the nature of the field margins.

The importance of nectar and pollen in the diet of hover flies has already been mentioned (page 34). The potential value of wild plants in the nutrition of various beneficial insects has been reviewed by van Emden (1965a). Parasitic wasps and tachinid flies as well as hover flies visit flowers, and greater numbers of parasitic wasps were trapped in a variety of crops when flowers were present. Many of these wasps favour umbelliferous weeds, but flower preferences may vary with species. An attempt to introduce a parasitic wasp into a new environment apparently failed through absence of the wild flowers which normally provided it with nectar essential to its diet and maturation. Various wild plants will also harbour predators such as capsid bugs and lacewings. Destruction of hedge bottom vegetation by weedkiller can greatly reduce populations of ground beetles (Pollard, 1966). While uncultivated land, hedgerows, etc., may contain wild hosts of pests, they will also favour beneficial insects; this subject, which is being studied by the Nature Conservancy, needs careful review before too much 'waste land' is cleared because it is thought to be unhygienic or uneconomic.

In 1969, 138 different pesticides were listed as Approved Products under the Agricultural Chemicals Approval Scheme. Experimental evidence of the effects of many of these on predators and parasites is lacking, but from what is known there can be little doubt that most of the insecticides and acaricides (of which 48 were listed), and some of the fungicides and herbicides, are toxic to some or most of them. In the following account only general indications of toxicity will be given and, unless otherwise stated, toxicities of pesticides will refer to applications at dosage rates normally used in the soil or on crops. It must be remembered that the effects of these chemicals on insects and mites can be complicated by such factors as differences in the effect of direct sprays and residual deposits, the persistence of the chemical and the dosage applied, differences between types of formulation, and the age of the insect. With insects having larval and pupal stages, the adults appear generally to be more susceptible than the larvae, and in some groups, e.g., hover flies and ladybirds, the larvae of some species show extraordinary tolerance. With insects having nymphal stages, such as capsid bugs, the young are usually more susceptible than the adults. Pesticides differ according to their chemical nature in the way they poison insects. Most of those in common use can affect insects both by contact and as stomach poisons, though a study of the toxicity of 61 pesticides to two species of parasitic wasp and two species of ladybird (Bartlett, 1966) showed that some of them, although toxic by contact, were less harmful as stomach poisons than others. This suggests that with some of the insecticides formulations or methods of application might be devised which would restrict their contact action, so that they would kill certain pests as stomach poisons while having little effect on predators and parasites. Most of the materials tested proved to be repellent to the insects when tasted, the more toxic chemicals usually being the more repellent, but since even a taste was fatal this repellence provided no protection for the insects.

Organochlorine Pesticides

These include aldrin, BHC, chlordane, DDT, dieldrin, endosulfan, endrin, heptachlor, TDE and toxaphene, which are persistent insecticides toxic to a wide range of insects. Also in this group are chlorbenside, dicofol, tetradifon and tetrasul, which are also persistent but mainly active against mites.

Aldrin and Dieldrin

Aldrin and dieldrin have been used for control of cabbage root fly and various soil pests, and to a smaller extent as foliar sprays, for example against blossom (pollen) beetles (*Meligethes*) species and cabbage seed weevil (*Ceutorhynchus assimilis* (Payk.)) on brassica seed crops. Following the recommendations of the Advisory Committee on Poisonous Substances used in Agriculture and Food Storage, their use was restricted after 1964 mainly to the control on commercial farms of certain pests (cabbage root fly, large narcissus fly, wireworms in potato land, strawberry seed beetle and, under certain conditions, wheat bulb fly and leatherjackets) for which no acceptable alternative chemicals were available. In the soil aldrin is gradually converted into dieldrin, which is very persistent since it takes 4–7 years for half of the original quantity of insecticide to disappear, and perhaps 20–30 years for 90 per cent of it to be lost (Wheatley, 1965). In a survey in 1961 of 15 arable fields in England in which aldrin had been used as an insurance against wireworm damage to potatoes at some time since 1954, dieldrin residues were found in all of them, at levels ranging from traces to nearly 1 lb per acre (Wheatley, Hardman and Strickland, 1962). Ten of the fields had residues of $\frac{1}{10}$ lb or more per acre.

Aldrin and dieldrin are particularly toxic to adults and larvae of butterflies and moths, beetles, wasps and flies. Much of the experimental work on their effects on predators has been with the ground beetles and rove beetles predatory on the young stages of cabbage root fly. As soil dwellers these beetles and their larvae would be exposed for long periods to residues of these insecticides, either from recent applications or from soil treatments in previous years. Small quantities well-distributed in the soil, as they are likely to become with repeated cultivations, may have greater adverse effects than larger amounts irregularly distributed. In experimental field plots average soil levels as low as 0·1 p.p.m. (parts per million) or less of dieldrin ($\frac{1}{10}$ lb or less per acre to a depth of 3–4 in.) killed the ground beetles *Bembidion lampros* and *Trechus quadristriatus*, and *Aleochara* species of rove beetle (Coaker, 1966). The larger ground beetles *Harpalus aeneus* and *Pterostichus melanarius* were found to be less sensitive, requiring dosages about three times as great to kill them, and were not affected by dieldrin levels below those toxic to cabbage root fly larvae. Dieldrin residues in the soil also reduced parasitism of cabbage root fly larvae and pupae by *Aleochara* species and the parasitic wasp *Idiomorpha rapae*, although concentrations above 0·2 p.p.m. were required to affect the latter.

The effects of aldrin and dieldrin on the general soil fauna have been studied at Rothamsted Experimental Station (Edwards and Jeffs, 1965). Aldrin was applied to newly ploughed old pasture and worked into the soil, and samples were examined at intervals during the following four years. Most of the mites, except predatory ones, were killed and numbers of most species of springtail were greatly decreased. Symphylids, root-feeding aphids and caterpillars were only slightly affected and numbers of eelworms and earthworms were unchanged, but almost all fly and beetle larvae were killed. In experiments on arable land where owing to cultivation there are usually fewer soil animals than in grassland, aldrin was not found to decrease greatly the numbers of insects, mites and related animals, and earthworms and eelworms were unaffected (Edwards, Dennis and Empson, 1967). One of the greatest effects was a reduction in populations of certain springtails. In terms of biomass, i.e., the mass of animals as distinct from numbers of individuals, Coleoptera and Diptera were the most affected; though more pest species—mostly belonging to these two orders—were killed than predators, or animals that break down organic debris or aerate the soil.

BHC

BHC is a mixture of chemical isomers of which the gamma isomer is the most valuable as an insecticide. Both the mixture (technical BHC) and the gamma isomer (gamma-BHC or lindane) are used as sprays for the control of such pests as aphids, capsid bugs, apple sucker, thrips, sawflies, etc. Gamma-BHC is used in the soil against wireworms and leatherjackets and as seed dressings for control of wireworms, turnip flea beetles (*Phyllotreta* species), wheat bulb fly and cabbage root fly. BHC is persistent but less so than DDT. In the soil half of it disappears in 15–20 weeks and 90 per cent in 2–3 years (Wheatley, 1965). It greatly reduces soil populations of mites and springtails (Sheals, 1956).

It is highly toxic to adults and young stages of the black-kneed capsid, *Anthocoris nemorum* and *Orius majusculus*, deposits having a residual effect which lasts for about two weeks (Collyer, 1953), and to larvae and adults of *Chrysopa carnea* (Bartlett, 1964a). It is toxic to adult and larval *Stethorus punctillum*, though with little residual effect, especially on the larvae, some of which also survived a direct spray (Collyer, 1953). Both adults and larvae of the seven spot ladybird are affected to some extent (Way, 1949). Though there is less experimental evidence than in the case of aldrin and dieldrin, there is little doubt that BHC is highly toxic to predatory ground beetles and rove beetles. *Oligota flavicornis* is very susceptible (Collyer, 1953).

Adults of parasitic wasps are severely affected (Way, 1949; Bartlett, 1963), though their larvae in the bodies of their hosts, e.g., *Aphelinus mali* (Hald.) in woolly aphid, are unharmed (Borg, 1952). Adults and larvae of predatory midges, *Aphidoletes* species, and of hover flies are also very susceptible (Way, 1949), and hover fly larvae are killed by eating aphids poisoned by BHC (Sol and Sanders, 1959).

Toxicity to predatory phytoseiid mites is moderate but deposits remain toxic for only a few days after spraying (Smith, Henneberry and Boswell, 1963; Bartlett, 1964b; Niemczyk, 1965).

Chlordane

Chlordane is restricted in its use in Britain to the control of ants, wasps and earthworms. It has a low toxicity to some ladybirds (Bartlett, 1963) but little is known of its effect on ground beetles or other soil-living insects.

Chlorbenside

Chlorbenside is an acaricide used for control of spider mites on top and soft fruit. It is toxic to *Typhlodromus pyri* (Sanford, 1967); other work has shown that it has only slight direct effect although it decreases the number of eggs laid by the mites (van de Vrie, 1962). It is moderately toxic to some *Phytoseiulus* species studied in California (Smith *et al.*, 1963), but this effect persists for only a few days. It appears to have little if any effect on insects.

DDT

DDT is widely used as a spray for leaf-eating pests such as beetles and caterpillars. This insecticide is very persistent; deposits on foliage persist for several weeks, while in the soil residues do not fall to half the amount originally applied until 5–8 years have elapsed, and for 90 per cent to be lost may take 25–40 years (Wheatley, 1965). Although not recommended as a soil insecticide, residues may accumulate in the soil from sprays applied to crops, for example fruit trees, peas and brassicas. In a survey in England in 1961 (Wheatley *et al.*, 1962), 14 out of 21 arable fields examined contained residues of DDT and its breakdown products ranging from traces to nearly 4 lb per acre, and in ten of the fields residues exceeded $\frac{1}{10}$ lb per acre.

DDT is more toxic than aldrin to predatory mites in the soil but does not harm springtails; in practice one of the most marked effects of DDT is an increase of springtail populations due to suppression of the predatory mites (Sheals, 1956; Edwards and Jeffs, 1965; Edwards *et al.*, 1967). This effect has occurred in experiments with both ploughed-up grassland and with arable land. On both types of land earthworms and eelworms were unaffected. In the ploughed grass fewer fly and beetle larvae were killed by DDT than by aldrin, but more symphylids and caterpillars. Expressed in terms of biomass, DDT had an equally depressing effect on pests and predators in the arable land study but in contrast to aldrin had little effect on beneficial animals breaking down organic debris. DDT will also reduce soil populations of millepedes and harvest spiders (Dempster, 1967).

Some of the predatory bugs on fruit trees are very susceptible to DDT, for example the black-kneed capsid (Gratwick, 1965). DDT is often used in June and July on apple to control codling moth, and one spray in July will virtually eliminate black-kneed capsid (Muir, 1966). An earlier spray at petal fall is also harmful and will greatly reduce populations of the capsid, while even at the green cluster stage DDT has some effect (Collyer, 1953). The black apple capsid and *Deraeocoris* species are also very susceptible (MacPhee and Sanford, 1956, 1961), although *Anthocoris nemorum*, especially the adult, is less affected (Gratwick, 1965). Advantage was taken of the comparatively low susceptibility to DDT of the anthocorid bug *A. musculus* (Say) in setting up a single species predator population in orchard experiments in Nova Scotia (Lord, 1962); successive sprays with DDT at low dosage rates resulted in a predator population consisting of practically 100 per cent *A. musculus*, so that its value could be studied without complicating effects from other predators. Other anthocorids, e.g., *Orius* species, may be more susceptible (Collyer, 1953).

DDT, as a deposit from a wettable powder spray, is highly toxic to adults of the green lacewing *Chrysopa carnea* but only slightly so to the larvae, and the eggs are unaffected by this insecticide (Bartlett, 1964a). Ladybird species vary in susceptibility. Adults and larvae of the seven spot ladybird are moderately susceptible to direct sprays or spray deposits, DDT emulsion having a greater effect than suspension sprays (Way, 1949), though the adults are less susceptible than those of the two spot ladybird (Way, Smith and Potter, 1954). Larvae of the two spot ladybird are severely affected by DDT emulsion deposits and moderately by those from wettable powder sprays, while adults and larvae of the 14 spot ladybird are only slightly affected by DDT (Gratwick, 1965). Species of *Stethorus* show considerable tolerance to DDT (Bartlett, 1963). Although there is less experimental evidence than there is with aldrin and dieldrin, there is little doubt that DDT is highly toxic to many ground beetles and rove beetles, though with ground beetles there are differences between species. Adults of the strawberry seed beetle are relatively resistant to residues of this insecticide in the soil though their larvae are very sensitive; adult *Trechus quadristriatus* also appear to be little affected, while *Bembidion* species are susceptible (Dempster, 1967).

Adult hover flies (*Syrphus ribesii*, *S. luniger*, *Scaeva pyrastri* and others) are very susceptible to DDT (Way, 1949) and sub-lethal exposure will inhibit egg laying, while the larvae are little affected by direct sprays or deposits at concentrations above those normally used. Larvae of predatory cecidomyiid midges, *Aphidoletes* species, are very susceptible.

A high susceptibility to DDT is also shown by adult parasitic wasps, e.g., *Diaeretiella rapae* (= *Aphidius brassicae* Marsh.) (Way, 1949), *Aphelinus mali* (Borg, 1952), *Aphytis mytilaspidis* (LeB.) and *Ascogaster quadridentatus* Wesm. (MacPhee and Sanford, 1954), and species of *Aphycus* (= *Metaphycus*). *Spalangia* and *Leptomastix* (Bartlett, 1963). Larvae of parasitic wasps inside the bodies of their hosts may be unaffected by DDT sprays, e.g., *Aphelinus mali* in the woolly aphid (Borg, 1952), and the degree of parasitism of aphids on glasshouse chrysanthemums by *Aphidius matricariae* was not reduced by weekly applications of DDT dust (Wyatt, 1966).

The effects of DDT on some species of predatory mite are moderate, e.g., *Typhlodromus pyri* (Sanford, 1967), and on others they are large, e.g., *Amblyseius hibisci* (Chant) (Bartlett, 1964b).

Dicofol

Dicofol is used for control of red spider mite under glass and fruit tree red spider mite on apple. *Phytoseiulus persimilis* is very susceptible to dicofol (Hussey and Parr, 1965). Populations of *Typhlodromus pyri* are reduced by dicofol sprays, and some other species of predatory mite are practically eliminated (Sanford, 1967; Sanford and Herbert, 1967).

Dicofol is moderately toxic to adult parasitic wasps, but has a low toxicity to some ladybirds (Bartlett, 1963) and little to the green lacewing *Chrysopa carnea* (Bartlett, 1964a) or to capsid bugs (MacPhee and Sanford, 1961), although populations of certain capsid bugs may be partially reduced (Sanford and Herbert, 1967). Some species of anthocorid are unaffected (Sanford and Herbert, 1967).

Endosulfan

Endosulfan is used for control of gall mite on black currant, strawberry mite on strawberry, and bulb scale mite on narcissus under glass; it has considerable persistence on foliage.

It is highly toxic to adult parasitic wasps and several ladybirds, including species of *Stethorus* (Bartlett, 1963), and larvae of hover flies, e.g., *Syrphus balteatus* and *S. ribesii* (Sol and Sanders, 1959). Larvae of parasitic wasps in the mummified bodies of cabbage aphid are little affected (Shorey, 1963).

It has little or no direct harmful effect on *Typhlodromus pyri*, though it decreases the mite's egg production for about three weeks after spraying (van de Vrie, 1962); it is moderately toxic to two other typhlodromid mites occurring on strawberry in California (Smith *et al.*, 1963).

Endrin

Endrin is another very persistent chemical. Its use is mainly restricted to control of gall mites on black currant and blackberry, and tarsonemid mites on strawberry and narcissus under glass. Information on its toxicity to beneficial insects is limited, but it is known to be highly toxic to several species of ladybird and parasitic wasp (Bartlett, 1963), and the larvae and adults of the green lacewing *Chrysopa carnea* (Bartlett, 1964a). Larvae of hover flies, e.g., *Syrphus balteatus* and *Scaeva pyrastri*, are moderately susceptible (Sol and Sanders, 1959). The susceptibility of predatory mites ranged from low to moderate for several species studied in California (Bartlett, 1964b).

Heptachlor

Heptachlor is restricted in its use in Britain to seed dressings for special purposes on wheat and sugar beet. In the soil it kills earthworms and enchytraeid worms and reduces populations of mites and springtails (Edwards and Arnold, 1966).

TDE ('Rhothane')

TDE, which is closely related to DDT, is mainly used for control of caterpillars on apple, caterpillars and weevils on strawberry, and some pests of carnation and chrysanthemum. Spray deposits are highly toxic to black-kneed capsid and *Anthocoris nemorum* (Gratwick, 1965), and to the adults of *Chrysopa carnea* but only slightly so to the larvae (Bartlett, 1964a); they are moderately toxic to adults of the 14 spot ladybird but have less effect on the larvae (Gratwick, 1965). Adult parasitic wasps (Bartlett, 1963) and some typhlodromid mites (Sanford, 1967) are highly susceptible. On *A. nemorum* and the 14 spot ladybird, direct sprays have less effect than spray deposits (Gratwick, 1965).

Tetradifon

Tetradifon is an acaricide used on top and soft fruit and several glasshouse crops for spider mite control. It has little effect on *Typhlodromus pyri* and some other species of predatory mite (Sanford, 1967; Sanford and Herbert, 1967). It appears to be harmless to insects, e.g., several species of capsid and anthocorid bugs (MacPhee and Sanford, 1961; Sanford and Herbert, 1967), the green lacewing *Chrysopa carnea* (Bartlett, 1964a), ladybirds and parasitic wasps (Bartlett, 1963).

Tetrasul

Tetrasul, which is closely related to tetradifon, is used on top fruit against spider mites, and it appears to be similar to tetradifon in its effects on predatory insects and mites. It is harmless to several species of capsid and anthocorid bugs, and predatory mites including *T. pyri* (Sanford and Herbert, 1967).

Toxaphene

Toxaphene is used in Britain only for control of earthworms in turf. It has a moderate toxicity to some ladybirds (Bartlett, 1963) but little is known of its effect on ground beetles and other soil-living insects.

CARBAMATES

Carbaryl

Carbaryl is the only insecticide in this group at present in commercial use in this country and is widely used on apples for control of caterpillars, especially those of codling and tortrix moths. It is persistent and spray deposits are highly toxic to black-kneed capsid and *Anthocoris nemorum* (Gratwick, 1965), and to several other capsid and anthocorid bugs (MacPhee and Sanford, 1961), though it has less effect on some species, e.g., black apple capsid and *Campylomma verbasci*. Larvae and adults of the green lacewing *Chrysopa carnea* and adult parasitic wasps are very susceptible (Bartlett, 1963, 1964a); so are larvae and adults of the 14 spot ladybird and larvae of the two spot ladybird (Gratwick, 1965), and adults of several other ladybirds (Bartlett, 1963).

Predatory mites, including *Typhlodromus pyri* and other species, are very susceptible (Bartlett, 1964b; Sanford, 1967; Smith *et al.*, 1963), though in the Netherlands only a small harmful effect, with reduced rate of egg production, was reported (van de Vrie, 1962).

Carbamate Herbicides

Little is known of the effects of carbamate herbicides (see also page 68).

ORGANOPHOSPHORUS PESTICIDES

There are over 20 compounds in this group and many are in common use on agricultural and horticultural crops. A few have short persistence, breaking down in a few hours or days after application, but the majority persist for two or three weeks or longer. Most have the ability to penetrate leaves, so that a deposit on the upper surface of a leaf can affect insects on the lower surface. Many are systemic, that is they can be absorbed by foliage or roots and then translocated in the plant's sap. Almost all can kill by contact as well as, in the case of the systemics, by poisoning insects or mites feeding on the sap. The majority are harmful to beneficial insects.

Azinphos-methyl

Azinphos-methyl is used for control of several top and soft fruit pests, pea moth (*Laspeyresia nigricana* (F.)) and cabbage root fly. It is persistent and highly toxic to black-kneed capsid and *Anthocoris nemorum* (Gratwick, 1965), and to several other species of predatory bug in Nova Scotia, though there some capsids and an anthocorid were found to be less susceptible or unaffected (MacPhee and Sandford, 1961). The adults and larvae of the green lacewing *Chrysopa carnea* are very susceptible (Bartlett, 1964a), so are adults and larvae of the 14 spot ladybird and larvae of the two spot ladybird (Gratwick, 1965), adults of several other ladybird species and parasitic wasps studied in California (Bartlett, 1963), and several typhlodromid mites including *Typhlodromus pyri* (Bartlett, 1964b; Sanford, 1967; Smith *et al.*, 1963). Azinphos-methyl mixed in the soil is considerably less toxic to predatory ground beetles than dieldrin (Mowat and Coaker, 1967).

Chlorfenvinphos

Chlorfenvinphos is employed for control of cabbage root fly, carrot fly and frit fly, as drenches or granular formulations; at the rates of application likely to be used it is virtually non-toxic to ground beetles (Mowat and Coaker, 1967). Mixed in the soil chlorfenvinphos causes a reduction in populations of parasitic mites and an increase in some springtails, but has only slight effects on other soil animals (Edwards and Thompson, 1967).

Demeton-S-methyl

Demeton-S-methyl is a persistent systemic pesticide widely used on many crops for control of aphids and spider mites. It was formerly marketed as demeton-methyl, butthe more pesticidal component, demeton-methyl, is now manufactured separately. It is very toxic as a direct spray or dried spray deposit to adults and nymphs of the red apple capsid and nymphs of *Anthocoris nemorum*, though adults of the latter are less susceptible (Gratwick, 1965). Anthocorids on beans sprayed with demeton-methyl were not completely eliminated and eggs hatched apparently normally several days later (Way, 1958). Larvae of the 14 spot ladybird are very susceptible, the adults less so (Gratwick, 1965). Adults of the 11 spot and two spot ladybirds are also very susceptible to direct sprays, while those of the seven spot show some tolerance (Way, 1958). Some ladybird eggs and larvae survived on sprayed bean plants, but only small numbers even when spray concentrations were reduced to one-third of the normal rate (Way, 1960). However, in plots of sugar beet sprayed twice, in May and June, numbers of ladybirds a month after the last spray equalled those in unsprayed plots; it was considered that the ladybird populations were more affected by destruction of their aphid prey than by poisoning by the insecticide (Heathcote, 1963).

Demeton-S-methyl probably also has adverse effects on predatory mites. *Phytoseiulus persimilis* can be killed by eating red spider mites containing traces of this chemical (Hussey and Parr, 1966).

Diazinon

Diazinon is mainly used on glasshouse crops, brassicas, carrots and lettuce. It is moderately toxic to some capsid bugs (MacPhee and Sanford, 1961), highly toxic to the green lacewing *Chrysopa carnea* (Bartlett, 1964a), moderately or highly toxic to some ladybirds and parasitic wasps (Bartlett, 1963) and to typhlodromid mites (Sanford, 1967; Smith *et al.*, 1963). In moist soil diazinon is as toxic to ground beetles as dieldrin, but less so in dry soil (Mowat and Coaker, 1967); as it has a much shorter persistence than dieldrin its long term effects may be less deleterious. It decreases the numbers of millepedes, centipedes, springtails and fly larvae in the soil in the year of application, but no effects are apparent in the following year (Edwards and Arnold, 1965; Edwards, Arnold and Thompson, 1966).

Dichlorvos

Dichlorvos has a very short persistence and is principally used on glasshouse and some outdoor vegetable crops. Little appears to be known about its effects on predators and parasites.

Dimethoate

Dimethoate is a moderately persistent systemic pesticide widely used for control of aphids and spider mites on many crops. It is very toxic to the red apple capsid, *Anthocoris nemorum*, larvae and adults of the 14 spot ladybird (Gratwick, 1965), to several other ladybird species and parasitic wasps (Bartlett, 1963), to adults of *Chrysopa carnea*, though the larvae are less susceptible (Bartlett, 1964a), and to typhlodromid mites (Sanford, 1967; Smith *et al.*, 1963). Predatory mites can be killed by eating red spider mites containing traces of dimethoate (McClanahan, 1967).

Disulfoton

Disulfoton is a persistent systemic insecticide formulated as granules for control of aphids on a variety of crops, and carrot fly. In the soil disulfoton decreases the numbers of mites, some springtail species, millepedes, thrips and fly larvae, though no effects are apparent in the year following that of the application (Edwards and Arnold, 1965; Edwards *et al.*, 1966). Its effects on parasites and predators on the foliage of crops appears to be imperfectly known.

Ethion

Ethion has been used for control of spider mites on apple, though in this country it is now only available as a seed dressing against wheat bulb fly. It has a moderate to high toxicity to ladybirds and parasitic wasps (Bartlett, 1963), and typhlodromid mites (Smith *et al.*, 1963). It is probably also harmful to other parasites and predators.

Formothion

Formothion is the first break-down chemical product of dimethoate in the plant; it has similar uses to dimethoate as a crop spray and probably has similar effects on beneficial insects.

Malathion

Malathion has a fairly short persistence and is widely used on many kinds of crop for control of several species of sucking insect, sawflies, raspberry beetle (*Byturus tomentosus* (Deg.)) and other pests. It is very toxic to several species of capsid and anthocorid (MacPhee and Sanford, 1961), larvae and adults of the green lacewing *Chrysopa carnea* (Bartlett, 1964a), several species of ladybird and parasitic wasp (Bartlett, 1963), typhlodromid mites (Sanford, 1967; Smith *et al.*, 1963), and hover fly larvae (Sol and Sanders, 1959). Hover fly larvae are also killed by eating aphids poisoned with malathion.

Mecarbam

Mecarbam is used as a spray on apples and pears against aphids and suckers, and also as a granular formulation for control of carrot fly and cabbage root fly. There is little information about its effect on parasites and predators.

Menazon

Menazon is a persistent systemic insecticide used for control of aphids on hops and some vegetables. It is toxic to adults and larvae of ladybirds and to hover fly larvae, but more so to aphids, so that when applied as a spray to beans at a dosage rate sufficient to kill aphids many hover fly eggs and ladybird larvae survive; thus it has potentially useful selective properties (Way, 1960, 1961; Jones, 1961). It is moderately toxic to some typhlodromid mites (Sanford, 1967). In the soil it appears to have minor effects on the fauna (Raw and Lofty, 1964; Raw, 1965).

Mevinphos

Mevinphos is a non-persistent systemic insecticide used for aphid and caterpillar control on a variety of fruit and vegetable crops. It is very toxic to young and adult red apple capsids and *Anthocoris nemorum* (Gratwick, 1965), larvae and adult *Chrysopa carnea* (Bartlett, 1964a), larvae and adults of the 14 spot ladybird (Gratwick, 1965) and other ladybird species (Way, 1960; Bartlett, 1963), parasitic wasps (Bartlett, 1963), hover fly larvae (Way, 1960), and some typhlodromid mites (Bartlett, 1964b). The short-lived effect of mevinphos would give predators not directly exposed to the spray a better chance of survival in a treated crop than persistent insecticides; hover fly eggs have been shown to survive a spray application (Way, 1961).

Morphothion

Morphothion is persistent and systemic, and used for aphid control on hops and chrysanthemums. Little appears to be known about its effects on useful insects and mites.

Oxydemeton-methyl

Oxydemeton-methyl, which is the first break-down chemical product of demeton-S-methyl in the plant, is a persistent systemic pesticide used for aphid and spider mite control on many fruit and vegetable crops. It is very toxic to typhlodromid mites (Sanford, 1967) and probably has similar effects to those of demeton-S-methyl on parasites and predatory insects.

Parathion

Parathion is persistent and is mainly used on glasshouse crops against aphids and other sucking insects, leaf miners and spider mites. It is highly toxic to the black-kneed capsid, sprayed foliage remaining toxic to the adults for at least ten days, and although the eggs are not affected newly hatched nymphs die for up to 20 days after spraying (Collyer, 1953). Spray deposits are also lethal to other capsid bugs and to anthocorid bugs for at least ten days. Many other capsid and anthocorid bugs are severely affected (MacPhee and Sanford, 1956, 1961; Redenz-Rüsch, 1959; Sanford and Herbert, 1967). Parathion is also lethal to adults of the two and seven spot ladybirds (Way, Smith and Potter, 1954), although the latter survived when placed on foliage two or three days after spraying (Collyer, 1953); *Stethorus punctillum* and *Oligota flavicornis* are also very susceptible, even to deposits ten days old. *Chrysopa carnea*, both larvae and adults, are very susceptible (Bartlett, 1964a), as are adult parasitic wasps (Bartlett, 1963), and the immature stages of *Aphelinus mali* in the mummified bodies of the woolly aphid (Evenhuis, 1959). Larvae of several species of hover fly are very susceptible and they can also be killed by eating aphids poisoned with parathion (Sol and Sanders, 1959). Typhlodromid mites are severely affected (Sanford, 1967; Smith *et al.*, 1963).

Parathion mixed in the soil decreases the numbers of earthworms, mites and springtails, though little effect is seen in the year following that of the chemical application (Edwards and Arnold, 1965; Edwards *et al.*, 1966).

Phenkapton

Phenkapton is a persistent acaricide used on apples for control of spider mites. Little is known of its effects on predators and parasites, though it appears to be principally acaricidal in action.

Phorate

Phorate, a systemic and persistent compound, is formulated as granules chiefly for aphid control on a variety of vegetable crops. In the soil it is very toxic to earthworms (Edwards and Thompson, 1967) and although evidence is limited it is probably harmful to predators and parasites. Predatory mites can be killed by eating spider mites which have fed on plants treated with phorate as a soil drench (McClanahan, 1967).

Phosalone

Phosalone is a persistent insecticide and acaricide used on apples and pears to control aphids, caterpillars and spider mites. Little is known about its effect on beneficial insects and mites.

Phosphamidon

Phosphamidon is a systemic and persistent compound used for control of spider mites, codling moth, apple sawfly (*Hoplocampa testudinea* (Klug)), pear sucker (*Psylla simulans* Förs.), and aphids on fruit and vegetable crops. It is highly toxic to adults and larvae of the green lacewing *Chrysopa carnea* (Bartlett, 1964a) and moderately or highly so to several ladybird species; parasitic wasps are very susceptible (Bartlett, 1963), though their larvae in mummified bodies of aphids may be little affected (Shorey, 1963). *Typhlodromus pyri* is very susceptible (Sanford, 1967).

Schradan

Schradan is a persistent systemic compound used for aphid and spider mite control on hops, strawberry, cucumber, some vegetables and ornamental crops. Unlike the other organophosphorous pesticides mentioned, it has limited contact action and has been claimed as a selective insecticide affecting sucking insects and mites while sparing predatory ladybirds and larvae of hover flies and midges (Ripper, Greenslade and Hartley, 1950). A low toxicity to larvae and adults of the green lacewing *Chrysopa carnea* and several ladybird species, but a moderate to high toxicity to parasitic wasps, is reported (Bartlett, 1963, 1964a). In experiments with sprayed field beans, schradan was found to be relatively harmless to *Anthocoris nemorum*, *Orius majusculus*, adults and larvae of ladybirds (including the two, seven, 11 and 14 spot species), and to larvae of hover flies (including *Scaeva pyrastri*, *Syrphus ribesii* and *S. luniger*) (Way, 1958).

Thionazin

Thionazin is a persistent nematicide and insecticide used for control of eelworms in tulip, narcissus and chrysanthemum, and cabbage root fly. In the soil it is very toxic to predatory ground beetles (Mowat and Coaker, 1967) and moderately so to earthworms (Edwards and Arnold, 1967). Predatory mites can be killed by eating spider mites which have absorbed traces of thionazin from treated plants (McClanahan, 1967).

Trichlorphon

Trichlorphon has a short persistence and is used against some caterpillar and fly pests on fruit and vegetables. It has a high toxicity to several species of hover fly larvae (Sol and Sanders, 1959) and some predatory mites (Bartlett, 1964b).

Vamidothion

Vamidothion is a persistent systemic pesticide used on apple, pear and hops for control of aphids, apple sawfly, pear sucker and spider mites. There is little information on its effects on beneficial insects and mites.

Mineral Oils

Tar oil is used as a winter wash on dormant fruit trees and bushes, and petroleum oil, generally in combination with DNOC, is used in the early spring when the buds are breaking; both are used to kill the eggs of various pests. A refined petroleum oil which can be applied to plants in leaf is used on some glasshouse crops.

Tar oil destroys *Anthocoris nemorum* hibernating on fruit trees (Massee and Steer, 1929); it also kills larvae of powdery lacewings but appears to have little effect on the eggs of black-kneed capsid (Collyer, 1953).

DNOC-petroleum also kills larvae of powdery lacewings (Collyer, 1953) and will considerably reduce the hatch of the eggs of black-kneed capsid (Muir, 1965). It is doubtless also harmful to hibernating typhlodromid mites.

The refined '*summer petroleum*' or '*white oil*' sprays in glasshouses are highly toxic to *Phytoseiulus persimilis* (Hussey and Parr, 1965).

Inorganic Insecticides

The principal inorganic compound used as a crop spray is *lead arsenate*—at one time almost the only weapon against leaf-eating pests and codling moth but now largely superseded by synthetic organic compounds. It is to some degree a selective insecticide inasmuch as it is a stomach poison killing insects feeding on sprayed foliage but not, or only to a small extent, those sucking plant sap. It shows almost no toxicity towards *Anthocoris nemorum* and the 14 spot ladybird, but spray deposits are moderately toxic to adult black-kneed capsid and larvae of the two spot ladybird (Gratwick, 1965). Lead arsenate is harmless to several predatory bugs studied in Nova Scotia (MacPhee and Sanford, 1961) and to *Typhlodromus pyri*, though it has some effect on several other species of predatory mite (Sanford, 1967). Some parasitic wasps are adversely affected (MacPhee and Sanford, 1954, 1956).

INSECTICIDES DERIVED FROM PLANTS

These include nicotine, derris, ryania and pyrethrum. Ryania is not used in Britain, and for many purposes the other three have been superseded by synthetic chemicals.

Nicotine

Nicotine is a non-persistent aphicide which is useful under glass, especially on chrysanthemum for controlling aphids which have become resistant to organophosphorus insecticides, but on field crops it has been largely displaced by more efficient organophosphorus compounds. It has only slight effects on black-kneed capsid (Collyer, 1953) and is harmless to the eggs and newly hatched larvae of the two and seven spot ladybirds, adult seven spot ladybirds, and hover fly larvae (Way, 1949). In the form of nicotine sulphate it has no, or only slight, effects on predatory capsid and anthocorid bugs (MacPhee and Sanford, 1961), none on adults and larvae of the green lacewing *Chrysopa carnea* (Bartlett, 1964a), and is only slightly toxic to parasitic wasps (Bartlett, 1963). Several species of predatory mite are unharmed (Sanford, 1967).

Derris

The active principle of this non-persistent insecticide, extracted from the roots of certain species of the derris plant, is rotenone; it has short persistence and is used for control of raspberry beetle and sawfly caterpillars and as a garden insecticide against caterpillars and aphids. It kills some newly hatched nymphs of black-kneed capsid, and has even greater adverse effects if mixed with petroleum oil (Collyer, 1953). It is toxic to the eggs of the two and seven spot ladybirds, to the larvae of the seven spot ladybird and to a smaller extent the adult beetles (Way, 1949). Adults of the green lacewing *Chrysopa carnea* are very susceptible but toxicity to the larvae is low (Bartlett, 1964a), while parasitic wasps are moderately or highly susceptible (Bartlett, 1963). Hover fly larvae are unaffected (Way, 1949). Some predatory mites are susceptible (Bartlett, 1964b).

Pyrethrum

Prepared from the flower heads of certain species of pyrethrum, especially *Chrysanthemum cinerariaefolium*, this has several insecticidal components known as pyrethrins. These decompose rapidly, especially in sunlight, so that the insecticide has brief persistence. It is chiefly used for control of fly pests of mushrooms and in preparations for control of house flies, and fleas and lice on domestic animals, etc. Information on its effects on parasites and predators is limited, but it is known to be toxic to the seven spot and two spot ladybirds (Way *et al.*, 1954) and some parasitic wasps (MacPhee and Sanford, 1954).

Ryania

This is extracted from the ground stem wood of *Ryania speciosa* and the insecticidal principle has been named ryanodine. It is moderately effective for the control of codling moth and is used in 'integrated control' programmes in several countries, but in Britain has not proved an acceptable material for control of this pest. Spray deposits are harmless to the nymphs of black-kneed capsid and have a low toxicity to nymphs and adults of *Anthocoris nemorum*, larvae and adults of the 14 spot ladybird, and larvae of the two spot ladybird (Gratwick, 1965). *Chrysopa carnea* larvae and adults are little affected (Bartlett, 1964a) as are parasitic wasps (Bartlett, 1963). Ryania is harmless to several species of predatory mite (Bartlett, 1964b; Sanford, 1967).

Inorganic Fungicides

Lime Sulphur

This is used to a limited extent at strengths ranging from 1–3 per cent for control of apple scab and apple mildew, and sometimes at strengths as high as 5 and 10 per cent against diseases on black currant and cane fruit. It is harmful to several beneficial insects and mites. Fresh deposits evolve hydrogen sulphide which is very toxic; deposits from a 2·5 per cent spray in a closed container are lethal to nymphs of red apple capsid (Reed, 1959). Harmful pockets of the gas could accumulate under still conditions in sheltered parts of the tree, such as partially folded young leaves and in blossom clusters. A deposit from 2·5 per cent lime sulphur is also toxic to nymphs of black-kneed capsid, but that from lime sulphur at strengths below 1·0 per cent is harmless to nymphs of either species (Muir, 1962). Lime sulphur as a concentrate spray is also harmful. Repeated lime sulphur sprays can considerably reduce populations of black-kneed capsid (Collyer, 1953) and will reduce the hatch of its winter eggs (Muir, 1965). Populations of *Typhlodromus pyri* are also severely diminished by lime sulphur sprays (Muir, 1965).

Sulphur

Sulphur, as colloidal or wettable powder formulations, is used against powdery mildews on various crops. Species of capsid bug differ in their susceptibility to sulphur sprays. Nymphs of black-kneed capsid and red apple capsid are only slightly affected (Muir, 1962); some other species are unharmed while others may be adversely affected to some extent (MacPhee and Sanford, 1954, 1956, 1961). The green lacewing *Chrysopa carnea* is little affected (Bartlett, 1964a), as are many ladybird species, but adult parasitic wasps are susceptible (Bartlett, 1963) and are also deterred by sulphur deposits (Flanders, 1942). Some predatory mites, e.g., *Typhlodromus pyri*, are slightly affected, while several other species are more susceptible (MacPhee and Sanford, 1961; Sanford, 1967).

Copper Fungicides

Copper fungicides, including Bordeaux mixture, appear to be harmless to many parasites and predators (MacPhee and Sanford, 1954, 1956, 1961).

Organic Fungicides

These include binapacryl and dinocap for powdery mildew control on apple; captan, dithianon, dodine, organomercury compounds, thiram and ziram for control of apple scab and some other diseases; fentin hydroxide, mancozeb, maneb, metiram, nabam and zineb for potato blight and other diseases; and quinomethionate (oxythioquinox) used against spider mites on apple, and powdery mildews and spider mites on gooseberry and some glasshouse crops. The known effects on useful insects and mites of some of these organic fungicides are given below; little appears to be known about the effects of the others.

Binapacryl

This is also used for spider mite control on apple and is moderately toxic to *Typhlodromus pyri* (Sanford, 1967).

Captan

Captan has little effect on capsid and anthocorid bugs (MacPhee and Sanford, 1961; Muir, 1962), the green lacewing *Chrysopa carnea* (Bartlett, 1964a), ladybirds and parasitic wasps (Bartlett, 1963), and the predatory mite *Typhlodromus pyri* (Muir, 1962).

Dinocap and *Dodine*

These have little effect on nymphs of black-kneed capsid and red apple capsid (Muir, 1962); dinocap will reduce populations of *T. pyri* but not sufficiently to prevent its effectiveness as a predator of fruit tree red spider mite (Muir, 1965).

Thiram

Thiram may reduce populations of *Typhlodromus* species (van de Vrie, 1962) but little seems to be known of its effects on useful insects.

Quinomethionate

This is highly toxic to *T. pyri* and several other predatory mites on fruit trees (Sanford, 1967), and to *Phytoseiulus persimilis* (Hussey and Parr, 1965).

Zineb

This is as harmless as captan to many predators and parasites (Bartlett, 1963, 1964a; MacPhee and Sanford, 1961; Muir, 1962). It may reduce populations of *Typhlodromus* species (Besemer, 1964).

HERBICIDES

Simazine

Widely used for weed control in some vegetable and soft fruit crops and in orchards, simazine considerably reduces the soil populations of fly and beetle larvae, springtails, mites, millepedes and enchytraeid worms, the effects lasting for four or five months (Edwards and Arnold, 1964, 1965).

MCPA, *Linuron* and *Tri-allate*

These appear to have no effect on the soil fauna. No effects from MCPA were apparent where this herbicide had been applied annually for nine years (Davis, 1966).

DNOC

This is harmful to insects and mites of all kinds.

Other Herbicides

Little appears to be known about the effects of most of the other herbicides used on crops.

References

Bartlett, B. R. (1963). The contact toxicity of some pesticide residues to hymenopterous parasites and coccinellid predators. *J. econ. Ent.* **56,** 694–8.

Bartlett, B. R. (1964a). Toxicity of some pesticides to eggs, larvae, and adults of the green lacewing, *Chrysopa carnea*. *J. econ. Ent.* **57,** 366–9.

Bartlett, B. R. (1964b). The toxicity of some pesticide residues to adult *Amblyseius hibisci*, with a compilation of the effects of pesticides upon phytoseiid mites. *J. econ. Ent.* **57,** 559–63.

Bartlett, B. R. (1966). Toxicity and acceptance of some pesticides fed to parasitic Hymenoptera and predatory coccinellids. *J. econ. Ent.* **59,** 1142–9.

Besemer, A. F. H. (1964). The available data on the effect of spray chemicals on useful arthropods in orchards. *Entomophaga* **9,** 263–9.

Borg, A. (1952). Undersökningar över *Aphelinus mali* Hald.—blodlusens (*Eriosoma lanigerum* Hausm.) speciella parasit. *Meddn St. VäxtskAnst.* no. 60.

Coaker, T. H. (1966). The effect of soil insecticides on the predators and parasites of the cabbage root fly (*Erioischia brassicae* (Bouché)) and on the subsequent damage caused by the pest. *Ann. appl. Biol.* **57,** 397–407.

Collyer, Elsie (1953). The effect of spraying materials on some predatory insects. *Rep. E. Malling Res. Stn for 1952*, 141–5.

Davis, B. N. K. (1966). Insecticide and herbicide residues in soil. (d) The long term effects of MCPA on soil micro-arthropods. *Rep. Monks Wood exp. Stn for 1960–1965*, 58.

Dempster, J. P. (1967). A study on the effects of DDT applications against *Pieris rapae* on the crop fauna. *Proc. 4th Br. Insecticide & Fungicide Conf., 1967*, **1,** 19–25.

Edwards, C. A. (1965). Some side-effects resulting from the use of persistent insecticides. *Ann. appl. Biol.* **55,** 329–31.

Edwards, C. A. and Arnold, Margaret K. (1964). The side-effects of toxic chemicals in the soil on arthropods and earthworms. In: Johnson, C. G. Entomology department. *Rep. Rothamsted exp. Stn for 1963*, 147–8.

Edwards, C. A. and Arnold, Margaret K. (1965). Effects of insecticides in the soil. In: Johnson, C. G. Entomology department. *Rep. Rothamsted exp. Stn for 1964*, 184.

Edwards, C. A. and Arnold, Margaret K. (1966). Effects of insecticides on soil fauna. In: Johnson, C. G. Entomology department. *Rep. Rothamsted exp. Stn for 1965*, 186.

EDWARDS, C. A. and ARNOLD, MARGARET K. (1967). Effects of insecticides on soil fauna. In: Johnson, C. G. Entomology department. *Rep. Rothamsted exp. Stn for 1966*, 195–6.

EDWARDS, C. A., ARNOLD, MARGARET K. and THOMPSON, A. R. (1966). Effects of insecticides on soil fauna. In: Johnson, C. G. Entomology department. *Rep. Rothamsted exp. Stn for 1965*, 187.

EDWARDS, C. A., DENNIS, E. B. and EMPSON, D. W. (1967). Pesticides and the soil fauna: effects of aldrin and DDT in an arable field. *Ann. appl. Biol.* **60,** 11–22.

EDWARDS, C. A. and JEFFS, K. A. (1965). The persistence of some insecticides in soil and their effects on soil animals. *Proc. 12th Int. Congr. Ent. London, 1964*, 559–60.

EDWARDS, C. A. and THOMPSON, A. R. (1967). Effects of insecticides on soil fauna. In: Johnson, C. G. Entomology department. *Rep. Rothamsted exp. Stn for 1966*, 196.

VAN EMDEN, H. F. (1965a). The role of uncultivated land in the biology of crop pests and beneficial insects. *Scient. Hort.* **17,** 121–36.

VAN EMDEN, H. F. (1965b). The effect of uncultivated land on the distribution of cabbage aphid (*Brevicoryne brassicae*) on an adjacent crop. *J. appl. Ecol.* **2,** 171–96.

EVENHUIS, H. H. (1959). Effect van insekticiden op de bloedluisparasiet *Aphelinus mali. Meded. Dir. Tuinb.*, **22,** 306–11.

FLANDERS, S. E. (1942). *Metaphycus helvolus*, an encyrtid parasite of the black scale. *J. econ. Ent.* **35,** 690–8.

GRATWICK, MARION (1965). Laboratory studies of the relative toxicities of orchard insecticides to predatory insects. *Rep. E. Malling Res. Stn for 1964*, 171–6.

HEATHCOTE, G. D. (1963). The effect of coccinellids on aphids infesting insecticide-treated sugar beet. *Pl. Path.* **12,** 80–3.

HUSSEY, N. W. and PARR, W. J. (1965). Glasshouse red spider mite. *Rep. Glasshouse Crops Res. Inst. for 1965*, 78–9.

HUSSEY, N. W. and PARR, W. J. (1966). Glasshouse red spider mite. *Rep. Glasshouse Crops Res. Inst. for 1964*, 87–8.

JONES, D. PRICE (1961). Menazon: development of a selective systemic aphicide. *Proc. Br. Insecticide & Fungicide Conf., 1961*, **2,** 433–40.

LORD, F. T. (1956). The influence of spray programs on the fauna of apple orchards in Nova Scotia. IX. Studies on means of altering predator populations. *Can. Ent.* **88,** 129–37.

LORD, F. T. (1962). The influence of spray programs on the fauna of apple orchards in Nova Scotia. XI. Effects of low dosages of DDT on predator populations. *Can. Ent.* **94,** 204–16.

MCCLANAHAN, R. J. (1967). Food-chain toxicity of systemic acaricides to predaceous mites. *Nature, Lond.* **215,** 1001.

MACPHEE, A. W. and SANFORD, K. H. (1954). The influence of spray programs on the fauna of apple orchards in Nova Scotia. VII. Effects on some beneficial arthropods. *Can. Ent.* **86,** 128–35.

MACPHEE, A. W. and SANFORD, K. H. (1956). The influence of spray programs on the fauna of apple orchards in Nova Scotia. X. Supplement to VII. Effects on some beneficial arthropods. *Can. Ent.* **88,** 631–4.

MacPhee, A. W. and Sanford, K. H. (1961). The influence of spray programs on the fauna of apple orchards in Nova Scotia. XII. Second supplement to VII. Effects on beneficial arthropods. *Can. Ent.* **93,** 671–3.

Massee, A. M. and Steer, W. (1929). Tar-distillate washes and red spider. *J. Minist. Agric. Fish.* **36,** 253–7.

Mowat, D. J. and Coaker, T. H. (1967). The toxicity of some soil insecticides to carabid predators of the cabbage root fly (*Erioischia brassicae* (Bouché)). *Ann. appl. Biol.* **59,** 349–54.

Muir, R. C. (1962). Further studies of the toxicities of fungicides to predatory mirids occurring in apple orchards. *Rep. E. Malling Res. Stn for 1961,* 109–12.

Muir, R. C. (1965). The effect of sprays on the fauna of apple trees. III. The influence of certain fungicides and petal-fall insecticides on the fruit tree red spider mite, *Panonychus ulmi* (Koch) and its predators. *Rep. E. Malling Res. Stn for 1964,* 167–70.

Muir, R. C. (1966). The effect of sprays on the fauna of apple trees. IV. The recolonization of orchard plots by the predatory mirid *Blepharidopterus angulatus* and its effect on populations of *Panonychus ulmi*. *J. appl. Ecol.* **3,** 269–76.

Niemczyk, E. (1965). The distribution of *Typhlodromus pyri* Scheut. (Acarina: Phytoseiidae) on apple trees in spring. *Rep. E. Malling Res. Stn for 1964,* 180–3.

Pollard, E. (1966). Background studies. VI. The invertebrate fauna of hedgerows at Ashton Wold. *Rep. Monks Wood exp. Stn for 1960–1965,* 44–6.

Raw, F. (1965). Current work on side-effects of soil-applied organophosphorus insecticides. *Ann. appl. Biol.* **55,** 342–3.

Raw, F. and Lofty, J. R. (1964). The side-effects of toxic chemicals in the soil on arthropods and earthworms. In: Johnson, C. G. Entomology department. *Rep. Rothamsted exp. Stn for 1963,* 148–9.

Redenz-Rüsch, I. (1959). Studies of the insect pest and predator fauna of an orchard in 'Bergische Land' and the influence of pesticides on this fauna. *Höfchenbr. Bayer PflSchutz-Nachr.* **12,** 169–256.

Reed, W. (1959). A preliminary study of the toxicity of lime-sulphur and other fungicides to *Psallus ambiguus* (Fall.) (Hemiptera-Heteroptera: Miridae). *Rep. E. Malling Res. Stn for 1958,* 127–30.

Ripper, W. E. (1956). Effect of pesticides on balance of arthropod populations. *A. Rev. Ent.* **1,** 403–38.

Ripper, W. E., Greenslade, R. M. and Hartley, G. S. (1950). A new systemic insecticide bis (bis dimethylamino phosphonous) anhydride. *Bull. ent. Res.* **40,** 481–501.

Sanford, K. H. (1967). The influence of spray programs on the fauna of apple orchards in Nova Scotia. XVII. Effects on some predacious mites. *Can. Ent.* **99,** 197–201.

Sanford, K. H. and Herbert, H. J. (1967). The influence of spray programs on the fauna of apple orchards in Nova Scotia. XVIII. Predator and prey populations in relation to miticides. *Can. Ent.* **99,** 689–96.

Sheals, J. G. (1956). Soil population studies. I. The effects of cultivation and treatment with insecticides. *Bull. ent. Res.* **47,** 803–22.

SHOREY, H. H. (1963). Differential toxicity of insecticides to the cabbage aphid and two associated entomophagous insect species. *J. econ. Ent.* **56,** 844–7.

SMITH, F. F., HENNEBERRY, T. J. and BOSWELL, A. L. (1963). The pesticide tolerance of *Typhlodromus fallacis* (Garman) and *Phytoseiulus persimilis* A. H. with some observations on the predator efficiency of *P. persimilis*. *J. econ. Ent.* **56,** 274–8.

SOL, R. and SANDERS, W. (1959). Über die Empfindlichkeit von Syrphiden-larven gegen Pflanzenschutzmittel. *Anz. Schädlingsk.* **32,** 169–72.

VAN DE VRIE, M. (1962). The influence of spray chemicals on predatory and phytophagous mites on apple trees in laboratory and field trials in the Netherlands. *Entomophaga* **7,** 243–50.

WAY, M. J. (1949). Laboratory experiments of the effect of DDT and BHC on certain aphidophagous insects and their hosts. *Bull. ent. Res.* **40,** 279–97.

WAY, M. J. (1958). Effects of demeton-methyl on some aphid predators. *Pl. Path.* **7,** 9–13.

WAY, M. J. (1960). Bean aphid (*Aphis fabae* Scop.) control. Effect of aphicides on beneficial insects. *Rep. Rothamsted exp. Stn for 1959*, 125–6.

WAY, M. J. (1961). Effects of insecticides on beneficial insects. In: Potter, C. Insecticides and fungicides department. *Rep. Rothamsted exp. Stn for 1960*, 148–9.

WAY, M. J., SMITH, P. M. and POTTER, C. (1954). Studies on the bean aphid (*Aphis fabae* Scop.) and its control on field beans. *Ann. appl. Biol.* **41,** 117–31.

WHEATLEY, G. A. (1965). The assessment and persistence of residues of organochlorine insecticides in soils and their uptake by crops. *Ann. appl. Biol.* **55,** 325–9.

WHEATLEY, G. A., HARDMAN, J. A. and STRICKLAND, A. H. (1962). Residues of chlorinated hydrocarbon insecticides in some farm soils in England. *Pl. Path.* **11,** 81–90.

WYATT, I. J. (1966). *Myzus persicae* on chrysanthemums. *Rep. Glasshouse Crops Res. Inst. for 1965*, 88–9.

POLLINATING INSECTS

BEES

The use of insecticides in agriculture and horticulture presents a potential hazard to bees as well as to other beneficial insects. Although effects of these chemicals are more difficult to assess on bumble bees and solitary bees than on honey bees, they must be considered because of the importance of wild bees as pollinators.

Beekeepers are apt to notice unusual occurrences in their hives, so information about poisoning of honey bees is comparatively easy to collect. Since 1956, honey bees alleged to have been poisoned have been reported to the Beekeeping Advisers of the National Agricultural Advisory Service and samples of these bees have been sent to Rothamsted Experimental Station to diagnose the cause of their death. In recent years most suspected poisonings involving substantial losses have probably been reported (Needham, Solly and Stevenson, 1966). Table 1, which summarizes the results of analyses for insecticide made on these bees, shows that the number of colonies affected is small in proportion to the total, but losses may be very serious for the individual beekeeper whose colonies are affected. Each incident may have involved any number of colonies, depending on the size of the apiary, and the number was not always reported.

TABLE 1

Insecticide poisoning of honey bees in England and Wales, 1959–67
(*From Needham and Stevenson,* 1966, *with additions*)

	Number of incidents			Approximate no. poisoned colonies	Total no. colonies	No. poisoned colonies per 1,000
	organo-phosphorus compounds	organo-chlorine compounds	others			
1959 ..	13	6	1	653	209,725	3·1
1960 ..	17	5	2	314	219,545	1·4
1961 ..	22	11	1	563	222,726	2·5
1962 ..	9	4	3	86	228,496	0·4
1963 ..	20	9	3	265	192,652	1·4
1964 ..	11	4	2	190	201,564	0·9
1965 ..	30	11	0	406	198,777	2·0
1966 ..	12	5	3	217	182,254	1·2
1967 ..	33	6	0	459	161,519	2·8

This work has shown that the most important cause of bees being poisoned is the spraying of field beans and, much less often, peas with organophosphorus insecticides to control aphids. Until now this has happened almost exclusively in eastern England but, with the acreage of field beans likely to increase in other parts of the country, it could become more widespread.

204 incidents of insecticide poisoning of honey bees were confirmed by analysis in the period 1959–66. Although evidence was not always available to suggest how the bees had acquired a lethal dose of insecticide, 34 per cent could be attributed to spraying of beans or peas, 5 per cent to the use of dieldrin spray to control blossom (pollen) beetles and weevils of mustard, and 4 per cent to dieldrin wasp baits. These last two causes should now have disappeared, although carbaryl, which has been substituted for dieldrin in some wasp baits, presents a slight hazard to bees in this formulation.

In addition to these three causes of poisoning, bees have occasionally been killed by other pest control operations such as the spraying of strawberries, kale, rape, top fruit, black currants and root crops. Carbaryl has been recommended as a fruit thinner for application to some varieties of apple from 80 per cent petal fall up to ten days afterwards, but fortunately is seldom used.

Bees are sometimes killed because containers are labelled inadequately, or people are ignorant of the implication of the labelling. For example, a wood preservative containing dieldrin was applied to hives; the label stated that it contained insecticide, but it was not stated that this made it harmful to bees. Gamma-BHC has been used to control was moth (*Galleria mellonella* (L.)) in combs by at least one beekeeper who did not realize it would kill his bees. On several occasions bees have been killed because captured swarms were put into sacks that had previously contained seed dressed with insecticide. Not all bee poisoning is accidental, for at least one swarm was deliberately sprayed with an insecticidal aerosol.

There is the possibility that systemic insecticides may render nectar or pollen toxic to pollinating insects, although present evidence does not suggest that this is a serious hazard.

Herbicides and fungicides rarely seem to kill honey bees in practice, although as will be seen from the list on page 74 some of them are potentially dangerous. Herbicides may have indirect effects. Thus, flowering weeds are sometimes important in tiding bees over periods when crops are not in flower, and spraying hedgerows with herbicides may greatly diminish the amount of forage available. Also, removing hedgerows may destroy nesting sites for wild bees over large areas, and for full yields of crops needing pollination it may be advisable to leave some nesting sites undisturbed.

The damage sustained by a colony of bees kept with'n flying range of a flowering crop that is sprayed with an insecticide is often largely confined to the loss of bees that are foraging on the crop when it is sprayed; the whole colony is not always lost. The proportion of bees affected depends on the attractiveness of the crop, on the flying distance from the hive to the crop, and on the weather. Feeding sugar syrup may enable a weakened colony to regain its original strength within a few weeks. However, the lack of a strong force of foragers for even a brief period in June or early July may not only mean the loss of a season's honey harvest for the beekeeper, but it also decreases the local population of pollinating insects when these could be of greatest use.

In many respects bumble bees and wild bees behave like honey bees, and they are probably poisoned under similar conditions. Consequently the following methods of protecting honey bees from insecticides should also protect other bees, and probably other pollinating insects.

Crops should only be sprayed with insecticide when they are not in flower and when any weeds they contain are not flowering, because bees will then not be foraging; few will be poisoned later by residual deposits unless very persistent insecticides have been used. In deciding when to spray orchards, it is particularly important to consider whether weeds on the orchard floor are flowering; the fact that bees also forage on weed flowers is often forgotten.

Few bees forage in the early morning or late evening and honey bees in particular are deterred by dull, overcast weather so that these times are most suitable for insecticide application. Foraging is also restricted by wind and rain, but it is unlikely that insecticides will be applied in such weather because of the danger that they will drift.

Insecticides can be applied to some crops in such a way that the bees will not come in contact with them, for example used as seed dressings or applied to the soil. Using certain organophosphorus insecticides on field beans as granules instead of as a spray also makes them very much safer to bees (Free, Needham, Racey and Stevenson, 1967).

When choosing the insecticide to use where bees may be at risk, the differences between the direct and residual toxicity should be considered, in addition to the safest method of application. The following list, issued by the Bee Advisory Service of the N.A.A.S. in 1968, and largely based on Johansen (1965), provides a valuable guide.

Toxicity of Agricultural Chemicals to Honey Bees

Products appearing in the Agricultural Chemicals Approval Scheme list for 1968 (other than those for use on hops and in glasshouses) may be classified as follows:

Insecticides

Group I: Highly toxic to bees working a crop (or flowering weeds among the crop) at time of treatment, and residually toxic for 24 hours or more

Aldrin	Dieldrin	Mevinphos
Azinphos-methyl	Dimethoate	Parathion
BHC	DNOC	Phosphamidon
Carbaryl	Formothion	Vamidothion
Diazinon	Lead arsenate	
Dichlorvos		

Group II: Toxic to bees working a crop at time of treatment, but not hazardous if applied when bees are not foraging

Chlorfenvinphos	Ethion	Odydemeton-methyl
DDT	Ethoate-methyl	Phenkapton
DDT with malathion	Malathion	Phosalone
Demeton-S-methyl	Mecarbam	TDE
Endosulfan	Menazon	Tetrasul
Endrin	Nicotine	Thionazin
		Trichlorphon

Group III: Non-hazardous to bees

Binapacryl	Disulfoton granules	Schradan*
Chlorbenside	Dicofol	Tetradifon
Derris	Phorate granules	

* Can appear in nectar and is stable in honey; possible hazard to consumer.

Fungicides

Non-toxic to bees, with possible exception of mercury compounds.

Herbicides

Dinitro compounds and endothal have a high toxicity to bees. Other herbicides are non-hazardous except when used on extensive stands of weeds (e.g., charlock) already in flower.

A large proportion of poisoning incidents involving field beans and other crops have followed spraying from aircraft. Drift on to neighbouring fields and inaccurate placing of insecticide are much more likely with aerial than with ground spraying application, and extra precautions are needed to avoid insecticide reaching crops where bees are foraging.

Damage to honey bees can be diminished by confining them in their hives while nearby crops are being sprayed, and in some parts of England spray warning schemes are operated to make this possible. However, delays to spraying caused by unfavourable weather, coupled with the limited period for which it is safe to confine bees, limit the value of the scheme; also a spray warning does not safeguard wild bees or other pollinating species.

References

FREE, J. B., NEEDHAM, P. H., RACEY, P. A. and STEVENSON, J. H. (1967). The effect on honeybee mortality of applying insecticides as sprays or granules to flowering field beans. *J. Sci. Fd Agric.* **18,** 133–8.

JOHANSEN, C. (1965). Bee poisoning. A hazard of applying agricultural chemicals. *Washington Agricultural Experiment Stations, Washington State University, Stations Circular* 356.

NEEDHAM, P. H., SOLLY, S. R. B. and STEVENSON, J. H. (1966). Damage to honey bee colonies (*Apis mellifera*) by insecticides in Great Britain, 1956–65. *J. Sci. Fd Agric.* **17,** 133–7.

NEEDHAM, P. H. and STEVENSON, J. H. (1966). Insecticides and beekeeping in England and Wales. *Bee World* **47,** 65–70.

Role of Predators and Parasites in Britain

THE controlling effect of predators and parasites has been closely studied for relatively few pests in Britain.

APHIDS

Aphids are attacked by many natural enemies, including larval and adult ladybirds (page 14), hover fly larvae (page 33), lacewings (page 10) and parasitic wasps (page 28). It is evident from field experience that, although these may give adequate control of aphids on some occasions, they cannot often be relied upon to provide the high degree of control usually required on crop plants, either to give clean produce or prevent virus spread. Many aphids, because their life cycles involve alternation between summer and winter host plants and further dispersal between summer host plants, are as it were a move ahead of the predators. The black bean aphid, for example, passes the winter as eggs on spindle bushes; these hatch early in the spring and the aphids feed on the foliage. As the new shoots mature in the early summer they become unsuitable as food, and winged aphids are produced which disperse to beans, beet, mangels and certain weeds. Further dispersal between these crops may occur during the summer, and in the autumn there is a return flight to spindle where the winter eggs are laid. When they arrive on beans in June, the aphids may be able to build up large colonies before predation has much effect, especially if there has been an abundant dispersal from spindle.

In studies at Rothamsted Experimental Station (Banks, 1955), ladybirds did not provide effective control of aphids in plots of beans; most of the ladybirds which had come out of hibernation were preying upon aphids on stinging nettles when black bean aphid began to arrive on the beans, and they did not spread to the beans in any numbers until the aphids on them had already become abundant. Of the other predators, lacewing larvae are known to be efficient, but although larvae of *Chrysopa carnea*, for example, killed about 95 per cent of black bean aphids on bean plants in one day in laboratory experiments, compared with 20–40 per cent killed by two spot ladybird larvae in six days (Banks, 1962), there are no estimates of their effects on aphid infestations on crops in Britain. Hover fly larvae also appear to be more efficient predators of black bean aphid than ladybird larvae (Banks, 1962). They are often the commonest predators of cabbage aphid, ladybirds only occurring infrequently on brassica crops, and were considered (George, 1957) to be the most efficient predators of this pest, but again their effectiveness in reducing aphid infestations has not been measured. Parasitic wasps appear to have a relatively small role. In some brassica plots studied in 1953 and 1954, the degree of parasitism of cabbage aphid by *Diaeretiella rapae* ranged from nil to about 7 per cent (George, 1957). On potatoes studied in 1944 and 1945, the percentage of aphids parasitized by *Aphidius* and *Praon* species varied from about 9–15, although on one crop it reached 50 in August and was over 90 in September (Dunn, 1949).

It is difficult to determine the effect of these various predators on aphid populations in the field. Observations were made in 1960 and 1961 on sugar beet plots at Broom's Barn and Rothamsted Experimental Stations, some of the plots being left unsprayed and some treated with insecticide (Heathcote, 1963). The insecticide treatments were a demeton-methyl spray on 19 May, or 15 June, or on both these dates, and phorate granules on 25 May. At Broom's Barn ladybirds, although common in neighbouring fields and hedgerows, did not become numerous on the young beet seedlings and could have had little effect on the colonizing black bean aphids. Aphids became sufficiently abundant to cause obvious damage on the untreated plants, though through the activities of predators the aphid infestations decreased in July. By contrast the treated plants became re-infested at this time and aphid populations increased, though not sufficiently to cause much damage. At Rothamsted where the beet was sown earlier, ladybirds were evidently active on the seedlings and killed many of the colonizing aphids. At this site aphid infestations never became large and caused little damage; by July the numbers of aphids on treated plants equalled those on the untreated.

The time at which predators invade a crop is probably important in deciding their influence on the growth of aphid colonies. In many cases use of insecticide is clearly necessary, though it is also important that predators should be allowed to prevent further increase in aphid infestations when the effects of insecticides have worn off, and to make their contribution to the decline in aphid numbers which normally occurs in August on sugar beet. In these experiments at Broom's Barn and Rothamsted the insecticides had little ultimate effect on ladybird numbers; the beetles were able to recolonize treated plants when aphid numbers began to recover, and in July, a month after the last spray, their populations on treated plants equalled those on the untreated. At Broom's Barn in 1961, after two demeton-methyl sprays in May and June, the population of adult ladybirds was estimated at 57,000 per acre. These included the ten and two spotted species, which were commonest, and the seven and 14 spotted species; other ladybirds were scarce.

Observations on the influence of predators on the fluctuations of black bean aphid populations on spring-sown beans were made over a nine-year period (1952–1960) at Rothamsted Experimental Station (Way, 1967). The aphid populations tended to alternate in successive years between large ones with peak densities between late June and mid-July and small ones reaching peak densities rather later, in early August. The large populations occurred when the majority of the bean plants were colonized by the primary migrant aphids from spindle; they were larger when the air temperature and sunshine were above average in June and early July, and smaller when weather over this period was cold and dull. A characteristic of the large populations was their rapid decline after the midsummer peak; this was considered to be due to 'delayed action' effects of predators coming from earlier-sown crops and wild host plants on which they had been multiplying. These would keep down aphid numbers in August and September, resulting in a small autumn flight to spindle and few eggs laid. There would consequently be few migrants leaving the bushes the following spring to colonize beans and weed host plants. With small infestations of aphids, predators would become scarce, so that aphids could multiply later in the summer on weeds and other crops such as beet and mangels. More migrants would become available for the return to spindle in the autumn, laying the foundation for large spring populations in the following year. Such a sequence would account for the two-year cycle.

Some weather conditions appeared to modify or compress this cycle. For example, in 1956 summer aphid populations were very small instead of large as expected. In the previous year exceptionally fine weather from July to mid-August enabled large populations of aphids to develop on late-sown beans and weeds; predators also increased, the aphid populations 'crashed' and few migrants reached spindle to found the 1956 colonies. In 1960 large aphid populations occurred instead of small ones as anticipated. This followed the long, fine summer of 1959, when in spite of predator pressure the aphids continued breeding; exceptionally warm weather from August to October probably favoured aphid multiplication when predators were beginning to hibernate. Consequently many aphids were able to migrate to spindle for winter egg laying, eggs were abundant and numerous winged aphids were produced in the spring of 1960.

Further work may be needed for a complete understanding of the factors determining the size of aphid populations before consistently reliable predictions can be made. At times other factors, such as weather conditions and fungal diseases, may have much greater effects than the natural enemies. Although the latter may not always provide sufficient control, their effect in checking the growth of aphid populations is nevertheless important and must be considered in relation to choice of insecticide. If an insecticide used for aphid control kills the natural enemies, it must be an efficient aphicide, otherwise there is a risk that surviving aphids will be able to multiply unchecked. In experiments on cabbage aphid control on Brussels sprouts in 1949 (Ripper, Greenslade and Hartley, 1950), when aphid infestations were severe, parathion sprays gave a good initial kill of aphids but also of predators, and the infestations built up again soon after spraying and more rapidly than on the unsprayed plants. Aphid resurgences on strawberries after using parathion are also reported. On beans, a DDT suspension spray which gave a partial control of aphids and killed the predators resulted in a crop yield no better than that from unsprayed control plots; increased yields resulted from a spray with an organophosphorus insecticide, which although it killed the predators was a good aphicide, and from a nicotine spray, which although inferior as an aphicide killed few of the predators (Way, Smith and Potter, 1954). Low-volume sprays of BHC-DDT applied to Bramley's Seedling apple trees at green cluster did not give good control of apple-grass aphid (*Rhopalosiphum insertum* (Wlk.)), and by petal fall infestations were higher than on the unsprayed control trees probably due to destruction of natural enemies (Moreton and Gould, 1960).

Problems of aphid control are also discussed in the section on 'Integrated Control', pages 103–8.

SMALL WHITE BUTTERFLY AND CABBAGE MOTH

The small white butterfly and the cabbage moth (*Mamestra brassicae* (L.)), both common and well-known pests of brassica crops, were chosen for a study at Monks Wood Experimental Station on the effects of pesticides on the balance between pests and their natural enemies (Dempster, 1966, 1967a, 1967b). A plot of Brussels sprouts was sprayed twice with DDT each year from 1964, the sprays being applied at the peak of each of the two generations of the butterfly; an adjoining plot was left unsprayed. Using a serological technique to identify the predatory insects and allied animals which had fed on the butterfly's young stages, it was shown that the predators attacking the young caterpillars included anthocorid bugs, several species of ground beetle, hover fly larvae, woodlice, spiders and harvest spiders. These regularly accounted for 50–60 per cent of the young caterpillars. Mortality also occurred towards the end of larval life and in the pupal stage, due to parasitism by *Apanteles rubecula* Marshall, *A. glomeratus* and two tachinid flies, *Zenillia* (=*Phryxe*) *vulgaris* (Fall.) and *Z. nemea* (Meig.), though this had a minor effect compared with that of the predators, accounting for about 10 per cent of the caterpillars. Predation by birds was also important. Natural mortality of the cabbage moth was also high in early larval life due to attacks by the same predators as attacked the young butterfly caterpillars.

The DDT sprays gave good control of the small white butterfly caterpillars present at the time of spraying but had little effect on the egg laying by the butterflies, so that, although the DDT killed caterpillars for several weeks, the pest could build up again later as the plants produced new leaves which were free of insecticide. At the same time the DDT killed many of the predators and this resulted in a marked reduction of the natural mortality of the second generation, following a DDT spray applied against the first generation. In each of three years of the study, 1964–1966, populations of spiders and anthocorids on the plants were reduced by about half, though hover fly larvae were less affected. These predators living on the plants did not, however, have so great a natural controlling effect as those hiding by day in the soil and surface litter and climbing the plants at night in search of prey. At Monks Wood the two most important predators not living on the plants were the strawberry seed beetle and the harvest spider *Phalangium opilio* L. The harvest spiders were severely affected by the proportion of DDT which reached the soil and, although they were able to recolonize the sprayed plot from adjoining unsprayed land, fewer tended to return each year as the insecticide residues accumulated in the soil. Numbers of adult strawberry seed beetles were not appreciably affected, probably because of recolonization since the number of larvae of this species in the soil decreased. Concentrations of DDT far below those required to kill the adults, however, were found greatly to decrease their rate of feeding so that their efficiency as predators was probably diminished. Populations of the ground beetles *Trechus quadristriatus* and *Nebria brevicollis* actually increased on the sprayed plot, probably because they are relatively resistant to DDT and because of the indirect effect of this insecticide in increasing the numbers of springtails (see page 56) which are their primary prey. Populations of species of *Bembidion* were reduced each year but they recolonized the plot. These various population trends may change as DDT residues in the soil accumulate to higher levels.

Of the parasites, *A. rubecula* was eliminated each year by the DDT, though *A. glomeratus* and the two tachinids were less affected. DDT appeared to increase the susceptibility of the caterpillars to granulosis virus disease as its incidence was higher on the treated plot.

Despite the toxicity of DDT to the caterpillars the effect of two annual sprays on the natural enemies was to increase the survival rate of the pest; in 1965, for example, there were twice as many caterpillars in August on the sprayed plants as on the unsprayed. It was even possible to show that DDT in the soil alone and not on the plants could result in a reduced natural mortality of the caterpillars. The experimental plot was small and adjoined uncultivated land containing large reserves of predators; in a more typically agricultural area recolonization of a treated area would probably be less marked, or even prevented if soil insecticide residues became too high. This work shows clearly the undesirable effects which can follow the use of a non-specific insecticide, especially one which is very persistent. Populations of cinnabar moth (*Tyria jacobaeae* (L.)) on ragwort on the edges of the same plot are also being studied to see if the effects of DDT on the predators and parasites extend beyond the treated area.

CABBAGE ROOT FLY

One of the most serious pests of brassica crops, this fly has two and sometimes three generations a year from April to September or October. It lays its eggs in and on the soil round the base of the stem of the plant, and the larvae hatching from them feed on the roots, causing damage which may result in stunting and sometimes death of the plant. Each female fly may lay about 80 eggs and if the flies are numerous about 500 eggs may be laid round a plant (Wright, Bowe and Buxton, 1955), so that there is a tremendous potential for root damage. It has already been mentioned (page 21), however, that over 95 per cent of the eggs of the first and second generations may fail to develop into adult flies. Although some of this mortality is on occasion due to desiccation of the eggs in dry soil conditions and, if larvae are numerous enough to cause severe root damage, some of them may fail to grow owing to death of the host plant, a high proportion of the mortality results from predation by the ground beetles and rove beetles described on pages 21–4.

The effect of this predation on the incidence of the pest and on crop damage was shown in experimental plots in 1958 and 1959 (Wright, Hughes and Worrall, 1960). Different population levels of predatory beetles during the period of activity of the first generation of cabbage root fly were obtained by surrounding some plots of early cauliflower plants with straw barriers treated with DDT to decrease the numbers of predatory beetles in the plots, while other plots were surrounded by barriers designed to make it easy for beetles to enter but difficult to leave, so that their numbers would increase. In the plots in which beetle numbers were decreased, there was a higher rate of survival of cabbage root fly eggs, more larvae and more root damage, resulting in a reduction in yield compared with plots from which beetles were not excluded.

From further experiments at Wellesbourne in 1960–62 (Coaker, 1965) using barriers to restrict movement of ground beetles in and out of brassica plots, it was estimated that ground beetles were responsible for up to one-third of the total egg mortality. The amount of predation varied according to the species and number of beetles present, and could also be affected by weather, especially heavy rain which breaks down soil crumbs and tends to cover and protect the cabbage root fly eggs (Hughes, 1959). *Bembidion lampros*, particularly active during the early stages of attack by the first generation of root fly in April and May, was probably the most important ground beetle at this time. In the later stages and during the second generation of the fly, species of *Pterostichus* and *Harpalus*, although less efficient predators individually, became common and thus more effective. Species of *Trechus* also became important as plant growth increasingly shaded the ground. The barriers used round the experimental plots did not exclude rove beetles, so that their predation might well explain a high proportion of the total root fly mortality not accounted for by predation by ground beetles.

Although the natural mortality of cabbage root fly is not always sufficient to prevent crop loss, which can be caused by comparatively few larvae, it has important implications in the use of insecticides. A reduction in predation as small as 10 per cent, through destruction of predatory beetles by insecticide, can double the number of root fly eggs surviving to produce larvae (Hughes, Salter and Mitchell, 1959). The more important predators have proved to be more sensitive than the pest itself to the insecticides which in the past ten years or so have chiefly been used against it. There are now numerous records from the National Vegetable Research Station of the effects of these insecticides on the predators. For example, in preliminary experiments in 1955 with plots in which aldrin, DDT and gamma-BHC at 2, 12 and $\frac{3}{4}$ lb per acre, respectively, were rotavated into the top three inches of soil, cabbages planted two weeks later were more severely damaged than those in untreated plots, as the insecticide levels in the soil were insufficient to control cabbage root fly but decreased the numbers of predators (Wright, Howells and Bowe, 1956). In 1963 with dieldrin levels of $\frac{1}{2}$–1 lb per acre in the soil, cauliflower yields were reduced by up to 60 per cent compared with those from untreated plots, due to destruction of predators and increased root fly attack (Coaker and Williams, 1964).

As species of *Bembidion* and *Aleochara*, which are the more susceptible, are the important predators over the period of the first cabbage root fly generation in May and June, their destruction by insecticide such as aldrin or dieldrin is likely to have striking effects at a time when root damage can be most severe, especially if aggravated by dry soil conditions. With crops exposed to the second generation of cabbage root fly and not the first, such as late-sown radish and swedes, dieldrin residues had less effect in increasing root damage (Coaker, 1966), probably because the chief predators at this time, species of *Harpalus* and *Pterostichus*, were less susceptible to low concentrations of aldrin and dieldrin in the soil.

While the use of some insecticide is necessary to ensure the degree of cabbage root fly control demanded in commercial crops, it is evident that the insecticide chosen must, if it also kills the predators, be fully effective against the pest. It is also desirable that, if a selective chemical cannot be found and one toxic to the predators must be used, it should be applied at the minimum dosage effective against the pest and to as small a proportion of the soil surface as possible, for example by using root dips or applying spot drenches to individual plants. The persistence of the insecticide should preferably not be greater than the period during which protection of the plant is required. A possible danger from the use of very persistent insecticides such as aldrin, dieldrin, DDT and gamma-BHC, whether used against cabbage root fly or other pests, is that in subsequent years residues toxic to predatory beetles will result in increased cabbage root fly attack, especially in brassica crops for stock feed not normally treated against this pest.

Another danger to be considered is development of resistance to insecticides by cabbage root fly. This was shown in 1963 to have occurred on a farm on the Cotswolds in Oxfordshire, where over 140 acres of Brussels sprout plants were destroyed or severely damaged, in spite of applications of aldrin. Aldrin and dieldrin have been the insecticides most widely recommended for control of this pest in recent years, though aldrin had been used on this farm at higher rates than necessary. The flies were found to be about 35 times more resistant to dieldrin, the breakdown product of aldrin in the soil, than flies from a farm where normal application rates still gave control, and they were also resistant to aldrin and gamma-BHC. None of the pupae collected was parasitized by *Aleochara* species or *Idiomorpha rapae* (Coaker, Mowat and Wheatley, 1963). Resistant flies apparently spread to neighbouring farms (Coaker and Mowat, 1965, 1966), and a survey showed that similar resistance had developed in many other parts of the country (Gostick and Baker, 1966).

Resistance to aldrin and dieldrin has arisen rapidly where these insecticides have been heavily used, due to the high selection pressure exerted on the fly populations and to the elimination of the natural enemies. The resistant flies are susceptible to organophosphorus insecticides, some of which can be used as a basis for alternative control measures. There is a risk that resistance to these insecticides may also develop in time, and considerable attention is being paid to their effects on the parasites and predators. The use of less persistent insecticides, with application rates kept to a minimum, may reduce the destruction of natural enemies and thus help to prevent or delay the selection of resistant flies. Similar problems may arise with other fly pests, such as bean seed fly (*Delia platura* (Meig.)) and wheat bulb fly.

FRUIT TREE RED SPIDER MITE

This is the classic example in Britain of a 'man-made' pest. It is a small red mite, occurring on many species of tree including fruit trees on which it passes the winter as bright red, spherical eggs laid on the bark. These hatch in May and June, and the females when mature lay eggs on the underside of the leaves; these summer eggs, as they are called, are also spherical but a lighter red. About a month after hatching, a new generation of mites is ready to lay eggs. Four or five generations may be produced during a favourable warm summer before the winter eggs are laid on the twigs and branches in September. The mites feed by sucking sap from the leaves and, if left unchecked, they can become so abundant that the foliage turns dull green and then bronze-coloured and may fall prematurely. The yield of fruit is also likely to be reduced, especially if the mites are abundant at the time of fruit set and development of the fruitlets.

Apple and plum are the fruit trees most likely to become infested and, especially on apple, sprays against fruit tree red spider mite are now a regular part of the annual spray programme. Yet before 1923 it was unknown as a pest. There can be little doubt from present knowledge that the rise of this mite to the status of a pest was due to the adoption in the 1920s and subsequently of new and more powerful insecticides than those previously available. The investigation by East Malling Research Station quoted earlier showed that, in an unsprayed orchard, fruit tree red spider mite was subject to predation by numerous species of capsid and other bugs (pages 4–8), several lacewings (pages 11–13), ladybirds (page 20), a rove beetle (page 24) and typhlodromid mites (pages 40–2). A glance at the life histories and habits of these predators will show that the chances of survival of the spider mite were slender. By the spring the proportion of viable winter eggs might on occasion be reduced by 40 per cent, and during the summer the onslaught by these various predators was intense and prolonged, so that the spider mite populations were kept down to a very low level.

It is easy to visualize the effects of insecticides on this balance between predators and a potential pest. Winter washes of tar oil and DNOC-petroleum, destroying lichens on the bark and anthocorid bugs and other predators sheltering amongst them, would enable spider mites hatching from winter eggs to start the season with considerable freedom from their enemies. It was with the widespread adoption of tar-oil winter washes in the 1920s, and later DNOC-petroleum, that the fruit tree red spider mite began to occur as a pest. The substitution of DDT and BHC sprays at the green cluster stage for winter washes and the use of BHC at petal fall, which became common practice in the 1950s and subsequently, also had drastic effects.

Some of the sequences of events in sprayed orchards have been studied (Collyer, 1953a, 1953b). These orchards were given winter oil washes, followed by fungicide programmes based on lime sulphur in the summer, and various insecticides. In contrast to the neglected orchards, the sprayed ones supported fewer species of predatory insect, although numbers of individuals of certain species might on occasion be very high. Populations of these predators fluctuated violently, varying from year to year and from orchard to orchard. In one orchard heavily infested with winter eggs of fruit tree red spider mite, a tar oil spray destroyed predators hibernating on the tree but had no effect on the mite eggs, so that large populations of the mite were able to develop. Black-kneed capsids came into the orchard in late July and August, and populations of other predators, especially ladybirds and anthocorid bugs, were also able to build up numbers by reinforcement from outside. These formed the basis of a large predator population in the following year, decreasing spider mite numbers so much that few winter eggs were laid. In another orchard, also heavily infested with spider mite eggs and sprayed with tar oil, the black-kneed capsid population was high enough to decrease the mites to such an extent in the ensuing summer that other predators failed to increase and some of the black-kneed capsids left for better feeding grounds. In a third orchard three parathion sprays were applied in June and July in an attempt to control fruit tree red spider mite chemically. This attempt had temporary success but killed all the predators, and three weeks after the last spray the few surviving mites began multiplying without check; leaf damage was evident in August, and in the autumn winter eggs were laid in such numbers as to redden the twigs and branches. An example is also quoted of the effects of DDT: trees sprayed with this insecticide on 1 July had 192 mites per leaf a month later compared with one per leaf on unsprayed trees.

At the time of these observations, lead arsenate—relatively harmless to the predators (see page 64)—was commonly used in June and July for control of codling moth. Fruit growers were advised to avoid using DDT after the green cluster stage, so as to give some of the predators a chance to survive, but during the 1950s DDT gradually became more widely used in the summer for codling and tortrix moth control as it was more effective than lead arsenate. The use of TDE, carbaryl, azinphos-methyl and other insecticides followed and, as these were all highly toxic to the predatory insects, the use of acaricides against fruit tree red spider mite became essential.

Although other differences between the neglected and cultivated orchards may play a part, such as altered micro-climate and reduced hibernation quarters on trees which are sprayed and pruned, it is evident that the spray programme has the major effect. This will range from partial reduction of predator populations by winter washes, followed by variable predator control of spider mite depending on the extent to which other harmful insecticides and fungicides are used in the summer, to almost total elimination of predators through using chemicals inimical to them several times during the season. At first sight the complex of prey and so many predators, and the effects of pesticides on them, might appear too complicated to manipulate in a predictable way, but interest in the possibilities of modifying spray programmes to allow biological control of fruit tree red spider mite led to closer investigation to measure the part played by a few of the important predators, namely the black-kneed capsid and typhlodromid mites.

The interaction of the black-kneed capsid and fruit tree red spider mite populations was studied over a seven-year period (1953–1959) at East Malling Research Station (Muir, 1965a, 1965b). Orchard plots of Cox's Orange Pippin and Worcester Pearmain apple were subjected to four different annual spray programmes over this period. Two of these included a DNOC-petroleum winter wash applied in March, followed in one programme by five sprays of lime sulphur and in the other by five sprays of captan. The third programme consisted of the lime sulphur sprays only, while the fourth consisted of captan only. To control the major pests apart from the mite, insecticides were chosen which would have the least adverse effects on the predator: nicotine was used against aphids and apple sawfly, and lead arsenate was applied pre-blossom for caterpillars and again in June for control of codling moth.

These programmes were chosen to give various population levels of the predator and pest so that their interaction could be followed over a succession of seasons. The winter wash was known to reduce the hatch of winter eggs of the mite, and lime sulphur to be toxic to some extent to the active stages, while captan was non-toxic. Both the winter wash and lime sulphur were toxic to the eggs of black-kneed capsid, while captan was harmless. Several species of predatory insect were present on the trees, but the black-kneed capsid was the most numerous, comprising on average over 90 per cent of the predator population.

The changes in population levels of fruit tree red spider mite will not be traced in detail here, but two marked fluctuations occurred, with highest numbers in the third and sixth years. The populations of the predator, which were low in the first year, increased and also showed two marked fluctuations, with the highest numbers in the fourth and seventh years, i.e., those years following the two peaks in mite numbers, and these were accompanied by drastic reductions in the mite infestations. Although the winter wash and lime sulphur treatments had some direct effect on the mites, this could provide no more than partial control, and the population changes which occurred were largely determined by the effects of these chemicals on the predator. Fruit tree red spider mite numbers were found to be inversely proportional to those of the predator, which were generally greatest on the captan plots and smallest on those sprayed with winter wash and lime sulphur. When mite infestations were greatest in their second upward surge in the sixth year, the numbers of active stages and eggs per leaf were as high as 220 on the winter wash + lime sulphur and winter wash + captan plots, compared with 85 on the lime sulphur only plots, and 20 on the captan only plots. There was a clear relationship between the rate of increase of the mite populations and predator numbers. It was calculated that the rate of increase in the absence of predation was 5·3 per cent per day, and that an average of 57 black-kneed capsids per tree was necessary to keep the mite numbers at a steady level. It was also shown that one predator could stabilize a population of about 2,000 mites and eggs, of which 40 per cent were active stages. With a lower predator to prey ratio than this the mites increased in numbers, but decreased with a higher ratio.

It was concluded from the pattern of mite and predator populations which became established on the plots that, under orchard management conditions which did not harm the predator, its numbers would reach a fluctuating equilibrium with those of the mite. Biological control of the mite could thus be obtained, provided that the mite infestations of the moderate levels which would at times occur could be shown to have no adverse effect on the growth and yield of the trees. Work on this aspect of the problem is in progress. In the above experiment black-kneed capsid also affected populations of leafhoppers, which were higher in the years when the predator was scarce than in those when it was numerous.

The role of the typhlodromid mites has been studied in several experiments. Different population levels of predatory mites and fruit tree red spider mites were achieved by subjecting plots of apple trees for three years (1951–53) to fungicide programmes with differing acaricidal effects—one based on lime sulphur, which is highly toxic to the predatory mites, two based on relatively harmless organic fungicides, and a fourth on dispersible sulphur, which is intermediate in effect (Collyer and Kirby, 1955). Populations of predatory mites increased in the first year on the organic fungicide plots and remained high, but remained low on the lime sulphur plots. Insect predators, unaffected by the fungicides, markedly reduced the initially high fruit tree red spider mite infestations, especially in June of the second year, and the latter then remained low on the organic fungicide plots but built up where lime sulphur was used, sufficiently to cause bronzing of the foliage in the third year.

In a further experiment (Collyer and Kirby, 1959) populations of fruit tree red spider mite remained substantially lower (seldom exceeding one per leaf) on apple trees subjected to a five-year (1952–56) fungicide programme based on captan than on trees sprayed with lime sulphur, which in the fourth year became bronzed because of the heavy infestations. Populations of *Typhlodromus pyri* were always higher on the captan-only trees and showed a marked inverse correlation with those of the spider mite, while the variable numbers of insect predators did not show a consistent correlation.

The results of these experiments, although not conclusive, suggest that typhlodromid mites are an important factor in the biological control of fruit tree red spider mite, though they could only have a decisive effect if the ratio of their numbers to those of the spider mite were relatively high, possibly not less than 1 : 10. An excess of predators over the pest until a considerable proportion of winter eggs had hatched in May seemed to be desirable if spider mites were to be prevented from increasing. Only on the organic fungicide plots was a sufficiently high ratio of predator to pest reached and maintained for two years. On the lime sulphur plots the ratio was always unfavourable; on the dispersible sulphur plots it was intermediate but definitely unfavourable in midsummer.

Experiments of this kind are complicated by the presence of other predators and possible unknown side-effects resulting from the use of chemicals. In another study (Collyer, 1958) known numbers of adult females of *Typhlodromus pyri* and fruit tree red spider mite were placed on seedlings of yellow bullace in an insectary. This represented quite a severe test as the introduced spider mite adults were able to make a good start since the predator prefers immature spider mites as prey. With predator/pest ratios ranging from 1 : 1 to 1 : 10 the spider mite numbers remained low throughout the three months of the experiment at less than one per leaf, or less than 50 per plant, compared with populations of some 3,000 mites per plant in the absence of the predator. With *T. finlandicus* the effect was similar, while three other species tested had little effect. It thus appeared that at least two species of typhlodromid mite could hold in check low infestations of fruit tree red spider mite and might play a useful complementary part to that of insect predators, such as black-kneed capsid, which are effective at higher spider mite densities.

Studies of the effects of modified spray programmes on black-kneed capsid and typhlodromid mites were further developed in experiments performed from 1961–1963 (Muir, 1965c). Captan used as routine sprays for control of apple scab is harmless to these predators, but it does not control apple mildew. Of the fungicides effective against mildew, lime sulphur was known to be toxic to the typhlodromid mites and young stages of the capsid, dispersible sulphur was less toxic to the mites and harmless to the capsid, while dinocap was harmless to the capsid and only moderately toxic to the predatory mites. Since the capsid does not hatch until early June, it might be possible to use lime sulphur safely before the blossom period. Of the insecticides, it had been found in the earlier work that nicotine at petal fall for apple sawfly did not interfere with the activity of the capsid, but the effect of gamma-BHC, commonly used at petal fall for sawfly control, was not known. The following experimental treatments were chosen:

(1) captan and dinocap;
(2) lime sulphur pre-blossom, dispersible sulphur post-blossom;
(3) captan only; and
(4) lime sulphur both pre- and post-blossom for comparison with the other treatments of its adverse effect on the predator.

Each fungicide programme was applied in combination with either nicotine or gamma-BHC at petal fall. All treatments included nicotine at green cluster for aphid control, and lead arsenate for summer sprays against codling moth. The fungicides were applied six times at fortnightly intervals beginning at the green cluster stage.

Over the three-year period of the experiment no adverse effects on populations of black-kneed capsid or *Typhlodromus pyri* by either the nicotine or gamma-BHC sprays at petal fall could be detected. There were, however, effects caused by the fungicides. Fruit tree red spider mite numbers tended to increase with both the lime sulphur treatments due to the toxicity of this fungicide to the two species of predator. Infestations actually became higher where dispersible sulphur was used, owing to the lower toxicity of this fungicide to the spider mite itself compared with lime sulphur; populations of black-kneed capsid also became higher on these plots in the summer because of the greater mite density compared with the other plots, but reduced hatch of the capsid eggs in the spring caused by the lime sulphur upset the predator-prey balance. With both the captan and captan + dinocap programmes, however, spider mite populations did not tend to increase, due to the activities of both predatory species, and by the third year of the experiment were not exceeding ten mites per leaf. Dinocap showed some toxicity towards *T. pyri*, but these mites are more effective with spider mite densities below ten per leaf, and sufficient survived to play their part.

The success of these experiments suggests possibilities of arranging an apple spray programme which would allow predator control of fruit tree red spider mite. There are, however, difficulties to be overcome, and the major one is control of codling and tortrix moths because these are controlled in the summer. Since lead arsenate or ryania do not give the high degree of control of codling moth provided by such insecticides as DDT, carbaryl or azinphos-methyl, all of which are lethal to the predators, either more selective chemicals or alternative means of control are required. It may be practicable to supplement the effect of the less efficient insecticides by measures directed against the overwintering larvae, such as corrugated paper bands placed round the tree trunks. Birds readily attack larvae in cocoons spun beneath the bands and may have a considerable effect in reducing codling moth populations. No alternative or supplementary control for tortrix moths can yet be suggested.

Although black-kneed capsid and typhlodromid mites can control the fruit tree red spider mite, it may be desirable to establish a more varied, and therefore more stable, fauna of predators. Since some of these predators are active in the spring, problems would arise in the use of insecticides at this time against the other major apple pests, which are not normally kept under sufficient control by natural enemies. These predators would also require alternative sources of prey, especially when spider mite populations were very low, so that the fruit grower might have to tolerate certain numbers of pest insects. More knowledge is needed of pest infestation levels which can be allowed to occur without causing an unacceptable degree of direct damage to the fruit or indirect losses by reducing growth or yield. Care in the choice of fungicides for disease control, especially those used frequently through the spring and summer, is obviously necessary.

With spray programmes modified for the benefit of predators, it may on occasion be necessary to apply one of the more toxic insecticides to deal with a pest outbreak. Therefore, the rate of recolonization of apple trees by black-kneed capsid after elimination by an insecticidal spray has been studied (Muir, 1966). Plots were sprayed once with DDT in late July at a time when such a spray might be applied against codling moth; almost all the black-kneed capsids present were killed. In the following year fruit tree red spider mite populations rose to a maximum of 100 mites and eggs per leaf, further increase being checked by the condition of the severely bronzed foliage. Black-kneed capsids entered the plots in the late summer but could not exert much effect as the ratio of their numbers to those of the mite was less than 1:4,000, considerably below the ratio of 1:800 needed to prevent mite increase (see page 84). Early in the third year, however, their numbers equalled those on adjoining control plots not sprayed with DDT, and exceeded them in the fourth year. Fruit tree red spider mite populations then fell to eight mites and eggs per leaf, the level which had been maintained throughout on the control trees. Control of the spider mite was thus obtained in three years. The plots were small, and with larger areas recolonization might take longer, especially if there are no sources of predatory bugs nearby. In this experiment leafhoppers also became more numerous in the second year in the DDT plots than in the untreated ones, but in subsequent years were less abundant in both treated and untreated plots due to predation by the capsid and also parasitism by *Aphelopus* species.

The role of other predators of fruit tree red spider mite has not been studied to the same extent as that of black-kneed capsid and typhlodromid mites, though it has been suggested that the role of *Anthocoris nemorum* is minor and transitory (Muir, 1966). Although this predatory bug feeds readily on the mite and tends to collect where this is plentiful, it tends to disperse when mite populations become low. Thus, because it emerges from hibernation before the winter eggs of the mite hatch, it is liable to leave the trees before the mites become available as food. In the investigation quoted above of the recolonization of apple trees after a July spray with DDT, it was found that in the following year *A. nemorum* entered the plots in numbers more or less equal to those of black-kneed capsid but did not persist and was scarce during the next two years.

Fruit tree red spider mite is one of the pests which has developed resistance to chemicals used for its control for several years. Although this problem can to some extent be overcome by changing to acaricides of different chemical groups, the number of such groups is at present limited. Development of new pesticides is very expensive and would become unnecessary if advantage could be taken of biological control.

RED SPIDER MITE

Also known as glasshouse red spider mite, as it is one of the major pests of many kinds of glasshouse crop, this mite can also occur at times as a pest of outdoor crops, such as hop, strawberry and black currant. It does not pass the winter in the egg stage as does fruit tree red spider mite but hibernates as adult females sheltering in the soil and plant debris. It seems more likely to reach pest numbers on soft fruit bushes in warm summers, and the risk of this happening is increased if predators are destroyed by spraying with insecticides such as DDT (Massee, 1954a).

It is not normally a pest of apple in this country, but progressively heavier infestations developed on trees in an Essex orchard repeatedly sprayed with parathion from 1949–1952 in an attempt to control fruit tree red spider mite (Collyer, 1953b). This provides another example of a 'man-made pest'. Increased infestations of this and other spider mites, including *Bryobia* species, on fruit trees following the use of DDT have been reported from many parts of the world.

Red spider mite resistance to organophosphorus insecticides on strawberry has occurred in several parts of Britain in recent years; this invites a study of the predatory typhlodromid mites found on strawberry and the effects on them of insecticides used against aphids.

References and Further Reading

BANKS, C. J. (1955). An ecological study of Coccinellidae (Col.) associated with *Aphis fabae* Scop. on *Vicia faba*. *Bull. ent. Res.* **46,** 561–87.

BANKS, C. J. (1962). Effects of the ant *Lasius niger* (L.) on insects preying on small populations of *Aphis fabae* Scop. on bean plants. *Ann. appl. Biol.* **50,** 669–79.

COAKER, T. H. (1965). Further experiments on the effect of beetle predators on the numbers of the cabbage root fly, *Erioischia brassicae* (Bouché), attacking brassica crops. *Ann. appl. Biol.* **56**, 7–20.

COAKER, T. H. (1966). The effect of soil insecticides on the predators and parasites of the cabbage root fly (*Erioischia brassicae* (Bouché)) and on the subsequent damage caused by the pest. *Ann. appl. Biol.* **57,** 397–407.

COAKER, T. H., MOWAT, D. J. and WHEATLEY, G. A. (1963). Insecticide resistance in the cabbage root fly in Britain. *Nature, Lond.* **200,** 664–5.

COAKER, T. H. and MOWAT, D. J. (1965). The dispersal of insecticide resistant strains of the cabbage root fly. *Rep. natn. Veg. Res. Stn for 1964,* 61.

COAKER, T. H. and MOWAT, D. J. (1966). The dispersal of dieldrin resistant cabbage root flies. *Rep. natn. Veg. Res. Stn for 1965,* 63.

COAKER, T. H. and WILLIAMS, DOROTHY A. (1964). Natural control of the cabbage root fly. *Rep. natn. Veg. Res. Stn for 1963,* 58.

COLLYER, ELSIE (1953a). The effect of spraying materials on some predatory insects. *Rep. E. Malling Res. Stn for 1952,* 141–5.

COLLYER, ELSIE (1953b). The greenhouse red spider mite, *Tetranychus urticae* Koch, on apples in East Anglia. *Rep. E. Malling Res. Stn for 1952,* 159–60.

COLLYER, ELSIE (1958). Some insectary experiments with predacious mites to determine their effect on the development of *Metatetranychus ulmi* (Koch) populations. *Entomologia exp. appl.* **1,** 138–46.

COLLYER, ELSIE and KIRBY, A. H. M. (1955). Some factors affecting the balance of phytophagous and predacious mites on apple in south-east England. *J. hort. Sci.* **30,** 97–108.

COLLYER, ELSIE and KIRBY, A. H. M. (1959). Further studies on the influence of fungicide sprays on the balance of phytophagous and predacious mites on apple in south-east England. *J. hort. Sci.* **34,** 39–50.

DEMPSTER, J. P. (1966). Interference in ecological systems by insecticides. b. The population dynamics of certain Lepidotera in the presence and absence of DDT. *Rep. Monks Wood exp. Stn for 1960–1965,* 60–1.

DEMPSTER, J. P. (1967a). A study of the effects of DDT applications against *Pieris rapae* on the crop fauna. *Proc. 4th Br. Insecticide & Fungicide Conf., 1967,* **1,** 19–25.

DEMPSTER, J. P. (1967b). The control of *Pieris rapae* with DDT. I. The natural mortality of the young stages of *Pieris. J. appl. Ecol.* **4,** 485–500.

DUNN, J. A. (1949). The parasites and predators of potato aphids. *Bull. ent. Res.* **40,** 97–122.

GEORGE, K. S. (1957). Preliminary investigations on the biology and ecology of the parasites and predators of *Brevicoryne brassicae* (L.). *Bull. ent. Res.* **48,** 619–29.

GOSTICK, K. G. and BAKER, PATRICIA M. (1966). Insecticide resistance in cabbage root fly. *Agriculture, Lond.* **73,** 70–4.

HEATHCOTE, G. D. (1963). The effect of coccinellids on aphids infesting insecticide-treated sugar beet. *Pl. Path.* **12,** 80–3.

HUGHES, R. D. (1959). The natural mortality of *Erioischia brassicae* (Bouché) (Diptera, Anthomyiidae) during the egg stage of the first generation. *J. Anim. Ecol.* **28,** 343–57.

HUGHES, R. D., SALTER, DOROTHY D. and MITCHELL, B. (1959). The natural control of cabbage root fly at Wellesbourne. *Rep. natn. Veg. Res. Stn for 1958,* 31.

MASSEE, A. M. (1953). Fluctuations in orchard fauna. *Rep. 13th Int. hort. Congr., London, 1952,* 250–5.

MASSEE, A. M. (1954). *The pests of fruits and hops.* London, Crosby Lockwood.

MASSEE, A. M. (1954a). Problems arising from the use of insecticides:

effect on the balance of animal populations. *Rep. 6th Commonw. ent. Conf.*, *1954*, 53–7.

MASSEE, A. M. (1958). The effect on the balance of arthropod populations in orchards arising from the unrestricted use of chemicals. *Proc. 10th Int. Congr. Ent., Montreal*, *1956*, **3**, 163–8.

MORETON, B. D. and GOULD, H. J. (1960). Small volume (concentrate) spraying of fruit trees. II. Experiments with insecticides. *Expl Hort.* **3**, 46–58.

MUIR, R. C. (1965a). The effect of sprays on the fauna of apple trees. I. The influence of winter wash, captan, and lime-sulphur on the interaction of populations of *Panonychus ulmi* (Koch) (Acarina: Tetranychidae) and its predator, *Blepharidopterus angulatus* (Fall.) (Heteroptera: Miridae). *J. appl. Ecol.* **2**, 31–41.

MUIR, R. C. (1965b). The effect of sprays on the fauna of apple trees. II. Some aspects of the interaction between populations of *Blepharidopterus angulatus* (Fall.) (Heteroptera: Miridae) and its prey, *Panonychus ulmi* (Koch) (Acarina: Tetranychidae). *J. appl. Ecol.* **2**, 43–7.

MUIR, R. C. (1965c). The effect of sprays on the fauna of apple trees. III. The influence of certain fungicides and petal-fall insecticides on the fruit tree red spider mite, *Panonychus ulmi* (Koch) and its predators. *Rep. E. Malling Res. Stn for 1964*, 167–70.

MUIR, R. C. (1966). The effect of sprays on the fauna of apple trees. IV. The recolonization of orchard plots by the predatory mirid *Blepharidopterus angulatus* and its effect on populations of *Panonychus ulmi*. *J. appl. Ecol.* **3**, 269–76.

RIPPER, W. E., GREENSLADE, R. M. and HARTLEY, G. S. (1950). A new systemic insecticide bis (bis dimethylamino phosphonous) anhydride. *Bull. ent. Res.* **40**, 481–501.

WAY, M. J. (1967). The nature and causes of annual fluctuations in numbers of *Aphis fabae* Scop. on field beans (*Vicia faba*). *Ann. appl. Biol.* **59**, 175–88.

WAY, M. J., SMITH, P. M. and POTTER, C. (1954). Studies on the bean aphid (*Aphis fabae* Scop.) and its control on field beans. *Ann. appl. Biol.* **41**, 117–31.

WRIGHT, D. W., BOWE, DOROTHY D. and BUXTON, J. (1955). Cabbage root fly. *Rep. natn. Veg. Res. Stn for 1954*, 53.

WRIGHT, D. W., HOWELLS, W. R. and BOWE, DOROTHY D. (1956). Natural enemies of the cabbage root fly. *Rep. natn. Veg. Res. Stn for 1955*, 47.

WRIGHT, D. W., HUGHES, R. D. and WORRALL, J. (1960). The effect of certain predators on the numbers of cabbage root fly (*Erioischia brassicae* (Bouché)) and on the subsequent damage caused by the pest. *Ann. appl. Biol.* **48**, 756–63.

Biological Control

THIS term is often used with different meanings and no single definition appears to be satisfactory. Paul DeBach (1964) proposed two definitions. The first, considering biological control as a phase of natural control, is that it is 'the action of parasites, predators or pathogens in maintaining another organism's population density at a lower average than would occur in their absence'. This definition does not imply any manipulation on the part of man nor whether the degree of control is economically satisfactory or not. The second takes into account the activities of man—'the study and utilization of parasites, predators and pathogens for the regulation of host population densities'.

Some of the known effects of predators and parasites contributing to natural biological control of British pests have been described on pages 75–88. In this section some examples will be given of biological control brought about by the deliberate introduction of natural enemies in various parts of the world. There are many successful ventures of this kind, mainly in connection with imported pests. With the exchange of goods between different countries of the world it was natural that insects should be taken to new lands, and their chances of making successful journeys were greatly increased as transport became more rapid, widespread and frequent. Many of the most serious pests in America, for example, originally came from Europe. Imported insects can become serious pests if they travel without the enemies which kept them in check at home and then find themselves in a suitable climate with the right food. The introduction of new crops or the growing of one kind of crop over large areas may give native insects greatly improved opportunities of reproduction.

BIOLOGICAL CONTROL OF PESTS OVERSEAS

The first international transfer of parasites was suggested in 1854 by Dr. Fitch, State Entomologist of New York, who was struck by the fact that a wheat midge accidentally introduced into the United States was seldom injurious in Europe where it was heavily parasitized. He attempted the importation of parasites but nothing came of it. The actual development of the idea of introducing parasites was mainly due to another American, Professor A. V. Riley. Co-operating with the Frenchman, J. E. Planchon, he attempted, although unsuccessfully, to control the vine louse or grape phylloxera (*Viteus vitifolii* (Fitch)), imported from the U.S.A. to France on vine rootstocks, by introducing a predatory mite from the U.S.A. in 1873. In 1874 the 11 spot ladybird and other natural enemies of aphids were taken from Britain to New Zealand, but there is little information about the results of this attempt. The parasite *Apanteles glomeratus* was introduced into the U.S.A. from Britain in 1883 to control the accidentally imported small white butterfly and it is still regarded as a useful member of the American fauna.

Fluted or Cottony Cushion Scale in California

Various other ventures in biological control were made, but all were eclipsed in 1889 by the spectacular results achieved by the introduction into California of an Australian ladybird, *Rodolia cardinalis* (Muls.), to combat the fluted or cottony cushion scale, *Icerya purchasi* Mask., which for some years had been causing such heavy losses in the orange and lemon groves as to lead to the abandonment of many of them. The scale was found to be a native of Australia so an entomologist was sent there by Riley to study the natural enemies. Since the ladybird was keeping the scale there well in check, a batch of ladybirds was sent to California where 129 arrived alive. The effects of their release were seen in the first season, and in two years the fluted scale menace had gone.

The ladybirds have since been introduced into many other citrus-growing countries with equal success. Their ability to control the scale insect so quickly is probably because they can complete one generation in a month compared with at least three months for a generation of the scale insect. Since they live only on the scale, or on one another when food becomes scarce, they are liable to die out when scale infestations have been reduced to a low level in any particular plantation; the scale can then multiply again but the ladybirds usually reappear soon, although deliberate reintroduction may occasionally be necessary.

Coconut Scale and Coconut Moth in Fiji

Another great success was the control in Fiji of the coconut scale, *Aspidiotus destructor* Sign. In Trinidad this insect does not appear to be a pest, being kept in check by a ladybird, *Cryptognatha nodiceps* Mshl., together with other natural enemies. Specimens of this ladybird were collected in Trinidad, taken to Fiji on scale-infested coconut palms and there reared on similarly infested banana leaves to increase their numbers. They were then liberated under coconut trees at various points in the Fiji Islands in 1928; within nine months the scale had been brought under control in all the more important islands, and by September 1929 it was rare in many localities. The continuance of this success in subsequent years showed that the ladybird was exercising a permanent control, which extended to many other host plants besides coconut. Unlike *Rodolia cardinalis* it could also prey upon another species of scale; this helped it to survive when it had cleared up the coconut scale. Its spectacular activities in Fiji, compared with those in Trinidad, were probably due to the absence of any enemies of its own, care having been taken not to introduce its parasites from Trinidad at the same time (Taylor, 1935).

In the case of the coconut moth, *Levuana iridescens* Beth.-Bak., another pest in Fiji, neither the original home of the insect nor any parasites were known. However, a fly parasitic on another moth attacking coconut, this time in Malaya, was collected without its hyperparasites and introduced into Fiji in 1923. By the following year it had established itself in all the islands of the group, exerting a control which appeared to be permanent. This was a remarkable instance of securing parasitism of an insect by a parasite normally attacking a different species, in this case not a closely related one.

European Spruce Sawfly in Canada

A great deal of work was undertaken in searching parts of Europe for parasites of the European spruce sawfly, which had been accidentally taken to Canada and was devastating spruce forests. The ichneumon parasite *Mesoleius aulicus* (Grav.), which was eventually found, gave control wherever it was introduced and established. But in 1948, nearly 40 years after the project was begun, the sawfly was reported as flaring up again in some areas and the rate of parasitism had fallen to a low level; for unknown reasons a large proportion of the parasite's eggs were failing to hatch. This illustrates the complexity of the subject of biological control (Baird, 1948).

Small White Butterfly in New Zealand

In 1929 the small white butterfly, well known in Britain breeding on brassica crops, appeared in New Zealand. By 1932 it had enormously increased in numbers and had spread over large areas; the whole of the North Island had been colonized by 1935 and the butterflies could be seen in millions. In 1932 and 1933 the parasite *Apanteles glomeratus* was imported from Britain, but the introduction proved a failure; this parasite has been successful in America and later introductions to New Zealand of the American strain gave better results.

The chalcid *Pteromalus puparum* was received from the Farnham House Laboratory in England and released in the summer of 1932–33; at the end of this season over 50 per cent of the butterfly pupae collected in a survey were found to be parasitized. Further releases were made in the following summer, and since then the parasite has provided a controlling factor of substantial importance (Cottier, 1948).

Woolly Aphid

The woolly aphid parasite, *Aphelinus mali*, has been introduced into many countries, including Great Britain (see also page 99). In Cape Province, South Africa, for example, it was introduced in 1920 from North America and gave good control of woolly aphid until suppressed by codling moth sprays, but soon became active again when these sprays were stopped (Georgala, 1953). In 1961 *A. mali* from Cape Province was introduced into orchards in Southern Rhodesia and produced remarkable results within six months; with the absence of codling moth and consequently the need to use insecticides toxic to the parasite, this biological method of control was considered to show great promise (Carnegie, 1962).

Citrus Mealybug under Glass

An interesting example of biological control of a glasshouse pest was the virtual elimination of a mealybug (*Planococcus citri* (Risso)) from stephanotis in a glasshouse in California by the introduction of an encyrtid parasite, *Leptomastix dactylopii* How., and the ladybird *Cryptolaemus montrouzieri* Muls. The parasites were introduced at the rate of 400 per plant and the ladybirds at one per plant, after the house had been closed in January and heated to maintain a temperature of 21°C (70°F) to force the plants into early flowering. The mealybugs were brought under complete control in three months, the percentage of infested plants falling from 66 in January to one in April and none in May (Doutt, 1952).

Winter Moth in Canada

This was first observed in 1949 in southern Nova Scotia and it subsequently spread to other areas as a serious pest of forest trees (Mesnil, 1967). Attempts at biological control were begun in 1952 and parasites were introduced from Europe. The principal parasite, the tachinid fly *Monochaeta* (=*Cyzenis*) *albicans* (Fall.), was established by 1955 and with the ichneumon *Agrypon flaveolatum* (Grav.) provided very successful control.

Further Examples

Many other attempts at biological control have been made. An outstanding success was the conquest of the Florida red scale, *Chrysomphalus aonidum* (L.), the major citrus pest in Israel, by importation from Hong Kong of the chalcid parasite *Aphytis holoxanthus* DeBach. Early establishment followed releases of shipments in 1956, and by 1959 control of the pest was complete (DeBach, 1964). Another serious citrus pest in Israel, the mealybug *Pseudococcus citriculus* Green, was controlled by the importation from Japan in 1940 of the parasite *Clausenia purpurea* Ishii. Following establishment of this parasite, the mealybug almost disappeared from many citrus groves (Rivnay, 1946).

Numerous other examples could be given. DeBach (1964) includes 111 pest species in his list of the cases of biological control of pests by imported predators and parasites. Of these, complete control of 21 was obtained in one country or another, and substantial control of 52. The majority are fruit or forest tree pests, and in many cases the attempts were made in only one or two countries. With certain pests, however, many countries have been concerned. The ladybird *Rodolia cardinalis* has been imported into 29 countries for control of the fluted scale, with complete success in 25. The parasite *Aphelinus mali* has been taken to 24 countries for control of woolly aphid, and complete or substantial control reported from 20. Altogether, considering the number of cases on a pest and country basis, they total 225, with 66 of complete control, 88 of substantial and 71 of partial control.

Taylor (1955) believed that biological control was the best of all control methods when it worked but that it did not work sufficiently often. He argued that it had its best prospects in tropical islands and had little future in continental areas. This view has been vigorously rebutted by other authorities and is not supported by DeBach's analysis, which shows almost equal numbers of successes for continental areas and islands and that successes have not occurred more frequently in tropical or sub-tropical than in temperate areas. It is interesting that about 40 per cent of the pest species controlled have been scale insects or mealybugs, and the majority of the remainder butterflies or moths, beetles, or plant bugs other than scales or mealybugs. Control by parasites has been obtained about four times as frequently as by predators. In general it is still difficult to account fully for successes in some cases and failures in others; much of the research effort has been guided by the seriousness of particular pests or the high value of the crops concerned, and there is still a tremendous field to be investigated and knowledge to be won.

BIOLOGICAL CONTROL OF PESTS IN BRITAIN

Red Spider Mites under Glass

Two species of red spider mite, *Tetranychus urticae* and *T. cinnabarinus* (Boisd.), are serious pests in glasshouses, attacking a wide range of crops including cucumber, tomato, French bean, peach, chrysanthemum, carnation, rose and many ornamental plants. The mites usually feed on the lower surface of the leaves and with heavy infestations the plants lose their healthy green colour and become sickly and stunted. *T. urticae* is the species usually found on cucumber, tomato and chrysanthemum; it also occurs on various outdoor crops. The mites hibernate as adult females in crevices in the glasshouse and then in the spring they lay eggs on the plants that are present. A succession of generations occurs from March to September.

This pest has frequently proved difficult and expensive to control, and the difficulty has been increased by the development of resistance by the mite to most of the available acaricides. A search for alternatives to the use of acaricides has included a study of the predatory mite *Phytoseiulus persimilis* (Plate VIII). This mite was obtained from Chile in 1959 by the German entomologist, G. Dosse, who subsequently supplied it to research workers in the Netherlands, Canada and the U.S.A. The adult mites are yellowish-orange in colour and are similar in body length to adult red spider mites but owing to their broader pear-shaped bodies appear larger. Both the oval eggs and the young stages are translucent and almost colourless. Young stages as well as adults move actively over the leaves in search of their prey, constantly waving and probing with their long front legs.

This mite was found to be a very efficient predator. Over a wide range of temperature its rate of development was twice as fast as that of red spider mite; at the optimum of 25–30°C (77–86°F) the female laid on average four eggs a day and during that time consumed an average of 30 eggs or 24 immature stages of red spider mite (Bravenboer and Dosse, 1962). In Canadian experiments with potted bean plants infested with red spider mite (Chant, 1961), it was found that the predatory mite, which appeared to depend almost entirely on animal food, showed a very efficient searching ability and established itself only in the presence of its prey. The females appeared to prefer adult red spider mites and, when they had eaten all that were available, moved off in search of new infestations of active mites, leaving the young hatching from their own eggs to feed on those hatching from eggs of their prey. This ensured wide dispersion of the predator and meant that its young would not hatch on leaves devoid of suitable food. At a predator/prey ratio of 1:50, red spider mites were virtually eliminated from the bean plants in three weeks. In the Netherlands it was found that the predator could eliminate red spider mite from peaches under glass before starving itself out, and on cucumbers could exert a control of the pest equivalent to that resulting from three to five sprays of acaricide (Bravenboer and Dosse, 1962).

In 1962 a series of experiments was begun in Britain to study the value of the predator as an alternative to chemical control and to find the most efficient way of using it (Hussey, Parr and Gould, 1965). In previous work (Hussey and Parr, 1963) on the effects of red spider mite infestations on yield of cucumbers, a scale of leaf damage categories was devised, ranging from 0–5 (1 = incipient damage; 5 = severe damage, leaf pale and beginning to shrivel with dense feeding marks over whole surface), where each step on the scale related to a certain number of active stages of mites per unit area of leaf. It was shown that the yield of cucumbers was not reduced until the mean damage category of the leaves reached about 1·9. It was therefore necessary to find at what leaf injury level and in what numbers the predator should be introduced to check the red spider mite infestations before plant damage became critical. The amount of leaf injury occurring in experiments, in which the predator was introduced at different red spider mite densities, showed a relationship between the density of infestations and the speed with which their increase was checked. The interval between introduction of the predator and the time when decrease of the mean leaf damage level occurred ranged from 21 days at low initial damage levels (i.e., low red spider mite infestation) to only ten days at relatively high initial levels. The number of predatory mites introduced varied from 2–14 per plant, and their ability for finding and consuming prey at high densities of the latter apparently more than compensated for the larger numbers of available prey.

At the beginning of an infestation there is only a slow increase in numbers of red spider mites, though later on unrestricted populations may increase the damage category by 1·0 within a week. Even with the maximum rate of increase of red spider mite ever observed, an initial mean leaf damage category of 0·2 should not exceed the critical level of 1·9 even with four weeks of unrestricted population growth. Therefore, taking a conservative estimate of four weeks for the predator to exert complete control, its introduction at a low initial leaf damage level of not more than about 0·2 would enable it to provide effective control.

The predator was found to persist for about eight weeks after introduction, so that fresh damage due to additional mites leaving hibernation quarters and reaching plants when the predator numbers had declined seemed unlikely before seven weeks after the original introduction. Fresh infestations could be controlled by reintroducing the predator when the leaf damage level had begun to rise again, in about the eleventh week.

In a further study (Parr and Hussey, 1967) with cucumbers which were artificially infested with red spider mites at planting, *Phytoseiulus persimilis* was introduced at rates varying from one on every plant to one on every second, fifth and tenth plants. Introduction on to every fifth plant was found to be the most economical and efficient way of using the predator.

The theories developed from these experiments were tested in large-scale trials in commercial cucumber houses (Gould, Hussey and Parr (in the press)). These trials confirmed that early introduction of the predator could maintain red spider mite populations at an extremely low level and often completely eliminate them. Where subsequent spider mite infestations occurred sporadically but regularly throughout the season, no further predator manipulation was necessary and complete control was maintained without chemicals. Usually, however, through destruction of its prey, the predator population declined to a very low level in June. This meant that spider mite infestations could develop subsequently from the few surviving spider mites accumulating on the youngest growth at the tips of the plants, or from mites entering the glasshouse from outside, and severe patches of damage could occur before the predator could regain control. To overcome this difficulty a system of 'banker' plants was proposed and this is described more fully below.

A provisional programme for the use of the predatory mite based on these studies has been suggested (Hussey, 1967). The size of the spider mite infestation in the spring will depend upon the number of mites going into hibernation in crevices in the glasshouse from the old crop in the previous autumn. This number can be greatly decreased by artificial lighting in the autumn to counteract the effect of shortening days and lengthening nights which induces the female mites to seek hibernation quarters; it will also be affected by continuation of predator control into the autumn. Consequently, the tactics to be adopted will differ in the years following that of the initial introduction of the predator.

In the first year, in glasshouses subject to severe spider mite infestations, it is suggested that to be on the safe side under commercial conditions two predators should be introduced on to every other plant within ten days of planting out, or before planting out if infestation occurs during the propagation stage. In either case the infestation should be made uniform by introducing spider mites, reared on French bean leaves, on to all plants that are not already infested. The most effective results are obtained by introducing the predator when the mean leaf damage category reaches 0·4; the predator–prey reaction thus set up should maintain a live predator population for at least eight weeks. Spider mites may continue to descend from winter quarters until May and these will be adequately controlled, at the same time extending the predators' period of activity at least until June.

To control infestations of spider mites which may occur after midsummer on the young growth at the tips of the plants, it is necessary to ensure that sufficient numbers of predatory mites are still available; as only small numbers from the original introduction will be present by this time a supplementary supply is required. It is suggested that one plant in every hundred (the 'banker') is infested with red spider mites which are allowed to increase until the plant is fairly severely damaged (leaf damage category =3·0) and the predator is then allowed access; at least 5,000 predators will then develop by the time control is regained. Portions of leaves bearing predators can then be distributed evenly on the other 99 plants. To prevent these banker plants becoming invaded by predators before the spider mite populations have become sufficiently large, it is suggested that low strength sprays of malathion (0·01 per cent, i.e., one-tenth of the commercial rate) should be applied to them weekly.

If production on these banker plants is staggered to provide maximum numbers of predators through July and August, effective control should be maintained with elimination of the potential overwintering population. The cost of banker plants in lost yield of cucumbers through red spider mite damage would be equivalent to about £10 per acre.

On nurseries subject to light infestations of red spider mite, the initial introduction of the predator should be made to only every fifth plant, following evening-up of the natural spider mite infestation.

In the second and subsequent years, owing to the control by the predator of the overwintering mites, infestation of the cucumber crops is unlikely to occur until midsummer, when mites dispersing from outdoor crops may enter the glasshouse. This situation could be met by a succession of banker plants to cover the period May to August. These plants should be selected and infested with spider mites eight weeks, and with the predator four weeks, before the time they would be required for use. An alternative course would be to inoculate all plants with spider mites in early June and introduce the predator to every fifth plant ten days later.

The potential value of biological control of red spider mite on cucumber under glass has thus been demonstrated, but large-scale production facilities will be required for its full realization. Relaxation of the use of insecticides for long periods may allow other pests such as *Aphis gossypii* Glover, glasshouse whitefly and thrips to become established, but there are good reasons for supposing that control of these by other natural enemies can also be achieved provided that a central production unit is available. The value of a parasitic wasp for biological control of whitefly is already known and this is discussed below.

In the meantime experience has shown that, although the predatory mite is susceptible to most insecticides used against other pests at standard application rates, many survive commercial high-volume sprays because of imperfect coverage of the leaves. Consequently, occasional sprays, especially with chemicals of short persistence, may be practicable. Dinocap has been used for control of mildew on cucumbers without detriment to the predator. Of the aphicides, nicotine, as a spray or fumigant, seems to be the only one which can be used without too serious an effect on the mite, although even this is not without risk. Otherwise smokes and fumigants in general are lethal and should be avoided.

Successful control of red spider mite by the introduction of *P. persimilis* has also been reported on tomatoes, chrysanthemums, peaches and orchids, but there have been failures on roses possibly due to adverse effects on the predator of sulphur vaporized for control of mildew, and to lower glasshouse temperatures (Hussey and Parr, 1966).

Glasshouse Whitefly

This small insect with white powdery wings is potentially one of the major glasshouse pests and can become very abundant if not controlled. During its scale-like nymphal stages, it attacks a wide range of plants and can be very damaging. Except in warm, sheltered situations it cannot survive our winter out of doors and was probably introduced some time ago on imported ornamental plants.

In 1926 a Hertfordshire gardener noticed that some whitefly 'scales' on his cucumber plants were black, and they were found by E. R. Speyer at Cheshunt Experimental and Research Station to contain parasites which were eventually identified as the chalcid *Encarsia formosa* (Plate IV) (Speyer, 1927). These insects were reared and distributed in tomato houses at the research station where they made good headway against whitefly infestations. In the following year about 10,000 parasitized whitefly scales were sent out to a dozen nurseries where the parasite became well established (Speyer, 1928). Very encouraging results were obtained in cucumber houses into which it was introduced in March and April; at the research station itself an early infestation of whitefly was completely eliminated by the end of April. Good control was also obtained in tomato houses. The parasite evidently spread to other nurseries for it was found in glasshouses, into which it had not been deliberately introduced, up to three miles from Cheshunt.

Following the building of a special glasshouse at Cheshunt for rearing large numbers of the parasite, it was distributed widely in Britain in 1928 and 1929 and a large consignment was sent to Canada to supplement the apparently weak stock of parasites existing there. Over two and a half million individuals were distributed in Canada in 1939 and 1940. In parts of South Australia the parasite reduced a serious whitefly problem to one of small proportions.

E. formosa is parthenogenetic and under warm glasshouse conditions it produces successive generations of individuals which are all females capable of parasitizing whitefly scales. It is winged and can fly over distances of at least several miles. The adult is extremely small, the size of a small pin's head, with dark head and thorax and yellow abdomen. It can live for about a fortnight and will die without laying eggs if whitefly scales are not available in that time. A single egg is laid in each scale, up to 50 scales being accounted for by one female; a fresh generation of adult parasites emerges after about a month, the insect leaving the empty skin of the scale by a circular hole on its back. About ten days after the egg hatches, the parasitized scale turns black, making it easily recognizable (Plate IV). Low temperatures cause a decrease in the number of eggs laid, and the parasite cannot survive the winter out of doors or in glasshouses kept for long below 11°C (52°F). At the high temperatures occurring in cucumber and hot houses, and in tomato houses in spring and summer conditions, it can eradicate severe whitefly infestations in two months.

Good control could not always be obtained on plants such as Nicotiana and the scented-leaved Pelargoniums, as the parasite seems averse to laying eggs in scales on sticky or aromatic leaves. Use was made of this fact in breeding the insect, by placing occasional tobacco plants in the rearing house as a means of preserving the whitefly stock. Tomato plants proved the most suitable for obtaining large numbers of parasitized scales on a small leaf area. The whitefly chooses the lower leaves for its egg laying; as soon as there are a sufficient number of black scales, the leaves can be removed for distribution, and one plant can provide a succession of leaves with parasitized scales from mid-April to autumn.

The breeding and distribution of *E. formosa* from Cheshunt Experimental Station was discontinued after the Second World War. As in the case of the woolly aphid parasite, its use was supplanted by that of modern insecticides; also it could not survive chemicals used against other glasshouse pests such as aphids and red spider mite. Another problem was the risk of accidental distribution of unwanted pests and diseases with the parasite material. Recent developments, however, with biological control of red spider mite under glass (page 94) have revived interest in *E. formosa* and, with the additional information which has been obtained on its biology and the interaction between it and the whitefly host, there are indications that it can be used for whitefly control on cucumber throughout the period when control of red spider mite by *Phytoseiulus persimilis* is in progress (Hussey, 1965).

The steady and controlled glasshouse environment offers possibilities of developing methods of biological control perhaps with more precision and predictability than can be expected on most outdoor crops. Other parasites in our fauna are being sought which in glasshouse climates may be more efficient than they normally are out of doors.

Woolly Aphid

The woolly aphid or American blight, a pest of apple trees, is believed to have been accidentally introduced from the eastern states of North America, probably on imported apple stocks. In 1922 the chalcid parasite *Aphelinus mali*, which had already been used successfully in New Zealand and Italy, was brought over. Introductions into Kent from 1924–37 did not appear successful, but after another attempt in 1938 strong colonies were found persisting in the orchards in the following year. Considerable numbers were also discovered in many Kent and Essex orchards in 1942 and 1943, and by the autumn of 1943 the parasites were evidently exercising a useful control of the woolly aphid.

It was known that the parasite could survive low winter temperatures, but it seemed likely that the wet conditions of the English winter and spring were not in its favour and probably made its establishment difficult and uncertain. It was, however, found that the parasite could be successfully overwintered in apple cold stores and released in orchards in the early summer, at a time later than that of natural emergence in the open, when weather conditions were more favourable and the aphids had begun breeding. Such a measure was adopted by many fruit growers. Shoots bearing colonies of parasitized aphids—the parasite overwinters as a larva or pupa within the skin of an aphid—were collected in October, freed of predators such as ladybird and hover fly larvae, and stored in apple boxes in a cold store at 4–8°C (39–46°F). Towards the end of May, in a spell of warm weather, the shoots were taken from the store and tied in small bunches to the branches of trees infested with woolly aphid, a few such trees being left unsprayed to allow the parasite to make a good start (Massee, 1954). With the advent of more effective insecticides for control of woolly aphid this procedure was abandoned; in any case the replacement of lead arsenate sprays for codling moth by contact insecticides such as DDT and azinphos-methyl was inimical to survival of the parasite.

WEED CONTROL BY INSECTS

Where plants accidentally or deliberately imported into foreign countries have become serious weeds, there have been many attempts to control them by the introduction of insects which feed on them. Probably one of the most spectacular examples was the campaign against the prickly pear cactus and other species of *Opuntia* in Australia. Several *Opuntia* species were taken to Australia between 1788 and about 1860 as hedge plants or as a source of fodder in low rainfall districts. Plants escaped from cultivation and became pests on a major scale in Queensland and New South Wales. In 1925 some 60 million acres were infested, so densely on half of this acreage that all grass and other herbage were choked out and the weight of the cactus growth was estimated to be 500–800 tons per acre. The land was useless for grazing or cultivation, and the dense thickets of prickly weed that spread for hundreds of miles presented a highly formidable clearance problem. Entomologists went to the American homes of the prickly pear, where they studied some 150 species of insect living on it; 60 of them were sent to Australia. From 1921 to 1924 two bugs, a moth, a red spider mite, and the cochineal insect *Dactylopius opuntiae* Cock. (=*D. tomentosus* Lamarck) were established; by 1926 the prickly pear acreage was decreasing instead of increasing. In 1925, eggs of the moth *Cactoblastis cactorum* (Berg), whose larvae bore into the cactus, were sent from Argentina, reared in cages, and during the next few years millions of eggs were distributed.

Within about two years the dense cactus cover had collapsed over several hundred acres; in 1930–32 vast areas were reduced to decaying pulp, and the last big tract was destroyed in 1933. In many areas fresh growth occurred, causing alarm, but was soon brought under control by the moth. Although the work took some ten years, the prickly pear territory was 'transformed as though by magic from a wilderness . . . to a scene of prosperous endeavour' (Dodd, 1936).

Similar work has been done with *Opuntia* species in other countries such as India and Ceylon; with St. John's wort in Australia, New Zealand, the United States and Canada; gorse in New Zealand and California; ragwort in New Zealand, Australia and the United States; and a number of others, all involving intensive programmes of search for and collection and study of potentially useful insects. All attempts have not yielded the desired results, but successes have been sufficiently frequent and valuable to give a great impetus to work on the biological control of weeds.

ORGANIZATION OF BIOLOGICAL CONTROL PROJECTS

The artificial employment of insect and mite predators and parasites for the control of pests, and their introduction into new countries, is obviously a skilled operation and one involving an immense amount of work. So many factors have to be carefully considered: the search for and identification of predators and parasites; their adaptability to a new environment; their efficiency in searching for their prey or host, speed of dispersal, and ability to maintain themselves when prey or host numbers become low; and in the case of insects for weed control, the likelihood of their attacking useful plants. Their life histories must be known and various factors affecting the relationship between predator and prey, or parasite and host, studied. Care must be taken to ensure that the predators and parasites themselves are free from parasites when they are imported into another country. Among many other problems there are those of safe transport, often over long distances, and of breeding to multiply stocks.

In 1927 the Imperial Institute of Entomology established in England the Farnham House Laboratory, with the object of investigating, on behalf of Dominion and Colonial entomologists, the natural enemies of introduced pests and of arranging the collection and transport of supplies. In the first nine years of its work the Laboratory studied 45 species and dispatched to various parts of the world over 600 shipments comprising over 11 million specimens of beneficial insects.

Owing to the war, work was interrupted at Farnham House in 1940 and operations were transferred to Canada. The organization was reconstituted as the Commonwealth Bureau of Biological Control in 1948 and moved its headquarters to Ottawa. The headquarters are now in Trinidad, and there are permanent laboratories in California, India, Switzerland and Pakistan. Other notable bodies concerned with the importation of useful insects as natural enemies include the Canadian Department of Agriculture's Institute of Biological Control, the United States Department of Agriculture's Insect Identification and Parasite Introduction Research Branch, and the University of California's Departments of Biological

Control at Berkeley and Riverside. Biological control is well established as a method of solving pest problems, whether alone or in combination with other methods, and is being pursued with increasing intensity.

References and Further Reading

ANON. (1947). White-fly parasite. (*Encarsia formosa* Gahan). *Experimental and Research Station, Cheshunt, Circular* No. 18.

BAIRD, A. B. (1948). Biological control. *Rep. 5th Commonw. ent. Conf., 1948*, 60–3.

BRAVENBOER, L. and DOSSE, G. (1962). *Phytoseiulus riegeli* Dosse als Prädator einiger Schadmilben aus der *Tetranychus urticae*-gruppe. *Entomologia exp. appl.* **5,** 291–304.

CARNEGIE, A. J. M. (1962). Woolly aphid of apple, *Eriosoma lanigerum* (Hsm.), and its control in Southern Rhodesia. *Bull. ent. Res.* **53,** 609–19.

CHANT, D. A. (1961). An experiment in biological control of *Tetranychus telarius* (L.) (Acarina: Tetranychidae) in a greenhouse using the predacious mite *Phytoseiulus persimilis* Athias-Henriot (Phytoseiidae). *Can. Ent.* **93,** 437–43.

COTTIER, W. (1948). Biological Control. *Rep. 5th Commonw. ent. Conf., 1948*, 63–5.

DEBACH, PAUL (Ed.) (1964). *Biological control of insect pests and weeds*. London, Chapman and Hall.

DEBACH, P. and LANDI, J. (1961). The introduced purple scale parasite, *Aphytis lepidosaphes* Compere, and a method of integrating chemical with biological control. *Hilgardia* **31,** 459–97.

DODD, A. P. (1936). The control and eradication of prickly-pear in Australia. *Bull. ent. Res.* **27,** 503–22.

DODD, A. P. (1940). *The biological campaign against prickly pear*. Brisbane, Commonwealth Prickly Pear Board.

DOUTT, R. L. (1952). Biological control of *Planococcus citri* on commercial greenhouse *Stephanotis*. *J. econ. Ent.* **45,** 343–4.

GEORGALA, M. B. (1953). The woolly aphis of apple and its control. *Fmg S. Afr.* **28,** 21–2, 27.

GOULD, H. J., HUSSEY, N. W. and PARR, W. J. (In the press). Large scale commercial control of *Tetranychus urticae* Koch on cucumbers by the predator *Phytoseiulus riegeli* Dosse. *Proc. 2nd Int. Congr. Acarology*. In press.

HUSSEY, N. W. (1965). Possibilities for integrated control of some glasshouse pests. *Ann. appl. Biol.* **56,** 347–50.

HUSSEY, N. W. (1967). Provisional programme for the use of the predatory mite *Phytoseiulus riegeli* to control red spider mite (*Tetranychus urticae*) on cucumbers. *Rep. Glasshouse Crops Res. Inst. for 1966*, 140–3.

HUSSEY, N. W. and PARR, W. J. (1963). The effect of glasshouse red spider mite (*Tetranychus urticae* Koch) on the yield of cucumbers. *J. hort. Sci.* **38,** 255–63.

HUSSEY, N. W. and PARR, W. J. (1965). Glasshouse red spider mite. *Rep. Glasshouse Crops Res. Inst. for 1964*, 78–9.

HUSSEY, N. W. and PARR, W. J. (1966). Glasshouse red spider mite. *Rep. Glasshouse Crops Res. Inst. for 1965*, 87–8.

HUSSEY, N. W., PARR, W. J. and GOULD, H. J. (1965). Observations on the control of *Tetranychus urticae* Koch on cucumbers by the predatory mite *Phytoseiulus riegeli* Dosse. *Entomologia exp. appl.* **8,** 271–81.

MASSEE, A. M. (1954). *The pests of fruits and hops*. London, Crosby Lockwood.

MESNIL, L. P. (1967). History of a success in biological control: the winter moth project in Canada. *Tech. Bull. Commonw. Inst. biol. Control* No. 8, 1–6.

ORDISH, GEORGE (1967). *Biological methods in crop pest control*. London, Constable.

PARR, W. J. and HUSSEY, N. W. (1967). Biological control of red spider mites on cucumbers: effects of different predator densities at introduction. *Rep. Glasshouse Crops Res. Inst. for 1966*, 135–9.

RIVNAY, E. (1946). The status of *Clausenia purpurea* Ishii and its competition with other parasites of *Pseudococcus comstocki* Kuw. in Palestine. *Bull. Soc. Fouad I. Ent.* **30,** 11–19.

SPEYER, E. R. (1927). White fly (*Trialeurodes vaporariorum*, Westw.). *Rep. Experimental and Research Station, Cheshunt, for 1926*, 55–6.

SPEYER, E. R. (1927a). An important parasite of the greenhouse white-fly. (*Trialeurodes vaporariorum*, Westwood). *Bull. ent. Res.* **17,** 301–8.

SPEYER, E. R. (1928). A white-fly parasite (*Encarsia formosa* Gahan). *Rep. Experimental and Research Station, Cheshunt, for 1927*, 75–7.

TAYLOR, T. H. C. (1935). The campaign against *Aspidiotus destructor*, Sign., in Fiji. *Bull. ent. Res.* **26,** 1–102.

TAYLOR, T. H. C. (1955). Biological control of insect pests. *Ann. appl. Biol.* **42,** 190–6.

Integration of Biological and Chemical Control

THE problems which have followed the use of modern insecticides, particularly that of the development of resistance to toxic chemicals by many pests, have stimulated an increase in work on more varied and subtle control measures, including combinations of chemical and biological control methods. The term 'integrated control' has been used by various entomologists with somewhat different meanings, but the definition adopted by the Food and Agriculture Organization (F.A.O.) Symposium on Integrated Pest Control at Rome in 1965 is 'a pest population management system that utilizes all suitable techniques in a compatible manner to reduce pest populations and maintain them at levels below those causing economic injury'.

These techniques may be of many different kinds including, for example, cultural methods such as crop rotations which prevent too great an increase in pest numbers at a particular site, or the use of crop varieties with resistance to certain pests; the use of lures based on sex attractant chemicals to trap insects, and chemosterilants to render pest insects sterile or other chemicals which disturb an insect's normal behaviour; and the use of disease organisms as well as predators and parasites. Natural enemies may be deliberately introduced, or those already present favoured by choosing selective chemicals which control the pest but spare at least some of its predators and parasites. The harmful effects of less selective chemicals on natural enemies may sometimes be minimized by applying them at a time when natural enemy activity is at a minimum, such as when the predators or parasites are in less susceptible stages in their life cycle (e.g., egg or pupa); by restricting the placement of chemicals, e.g., by using seed dressings, soil drenches or baits; by reducing the area treated, leaving strips or plots to maintain reservoirs of natural enemies; by decreasing the dosage level to give partial but adequate control of the pest while reducing the adverse effects on its natural enemies; and in the case of sucking insects by using systemic insecticides which have no contact action.

It is important to realize, as Kennedy (1968) has emphasized, that any single, effective and often applied method of pest control—not only the use of pesticides—will exert a selection pressure to which the pest may respond by developing resistance. Consequently we need to think in terms of control agents varied in kind and time of application, each mild in itself but in combination with others sufficiently effective for the time being. In other words we must try to imitate just enough of the diversity of selection pressures seen acting in Nature to relieve the pressure from any one agent, and thus prevent the pest getting out of control by developing resistance faster than we can revise our control measures (Smallman, 1965).

Attempts have already been made at modifying chemical spray programmes to favour the natural control of certain pests by predators and parasites. One of the pioneer studies began in Nova Scotia in the early 1940s (Ross, 1953; Dicker and Kirby, 1955). The loss of the British market for apples in 1940 had led to a depression in the apple-growing industry, so that numerous unsprayed orchards were available for study and growers were anxious to restore their trees to health at a low cost. A. D. Pickett was convinced that insecticidal spray programmes could be reduced if natural enemies were allowed to play their part. He showed that heavy infestations of mussel scale (*Lepidosaphes ulmi* (L.)), which were causing very serious damage, were due to destruction of a chalcid parasite (*Aphytis mytilaspidis*) and a predacious mite (*Hemisarcoptes malus* (Shimer)) by sulphur sprays used for control of apple scab. Substitution of copper and ferbam fungicides for sulphur quickly resulted in a sensational decline of mussel scale to a pest of minor importance.

It was realized that sulphur, especially lime sulphur, was also harmful to typhlodromid mites preying upon fruit tree red spider mite, and to a predatory thrips (*Haplothrips* sp.) which attacks the mite and eats the eggs of codling moth and bud moth (*Spilonota ocellana* (Schiff.)). With the high rainfall, scab is a serious problem, and copper and ferbam sprays were not entirely satisfactory, but these were followed by glyodin and then captan, which were effective fungicides and harmless to the predators and parasites. Insecticide usage was largely restricted to compounds having the least effect on predators and parasites, such as nicotine sulphate for aphids, lead arsenate for caterpillars, and ryania for codling moth. More widely toxic insecticides such as DDT were occasionally used for certain pests when the situation called for more drastic action, but usually at reduced dosages or at times when the natural enemies would be least harmed.

By 1965 integrated control programmes on these lines had been adopted on over 80 per cent of the apple and pear acreage in Nova Scotia (Pickett and MacPhee, 1965; Patterson, 1966). Surveys suggested that as the acreage managed in this way increased, damage by pests decreased, and high quality fruit was being produced at a cost per bushel less than half of that incurred with a full insecticide programme. Promising results with modified spray programmes were obtained elsewhere, for example in Wisconsin (Oatman, 1966). Various combinations of insecticides were tried on plots of apples over a period of three years from 1960–1962; a programme in which lead arsenate and ryania were the only insecticides used proved the most satisfactory in controlling the chief pests and allowing natural control of spider mites. In this trial and those in Nova Scotia the proportion of fruit damaged by pests, which included others besides codling moth, varied from about 5 per cent to levels sometimes over 20 per cent, a degree of damage which might not be acceptable to British growers. In other areas, for example British Columbia and the Niagara Peninsula, less success was experienced (Putman, 1963); a warmer climate enabled codling moth to breed more abundantly and the indigenous natural controls were unable to prevent serious fruit injury. Repeated sprays of ryania were not only too expensive but had adverse effects on some of the predators and parasites.

Although many of the major pests of apple are common to Nova Scotia and Britain, and the fauna of natural enemies is very similar, we do not possess the predacious thrips. In Nova Scotia, too, between fruit-growing farms there are large areas of rough country which probably harbour large populations of predators and parasites. Another difference is that apple mildew is almost unknown in Nova Scotia, and available routine fungicides used for its control in England are toxic to predatory mites. It has already been described (pages 82–88) how fruit tree red spider mite could be controlled by natural enemies but, apart from the need for great care in the choice of fungicides for mildew control, there is still the problem of codling and tortrix moths.

Another interesting story of integrated control is that of the spotted alfalfa aphid, *Therioaphis maculata* (Buckt.), which was found in California in 1954 (Stern, van den Bosch and Born, 1958; Stern, Smith, van den Bosch and Hagen, 1959; Stern and van den Bosch, 1959; Smith and Hagen, 1959; van den Bosch *et al.*, 1959). By 1956, thousands of acres of lucerne (alfalfa) had been sprayed with organophosphorus insecticides against this aphid, and in the autumn of that year failure to obtain control was reported from several areas. Tests showed that a low degree of resistance to these insecticides had developed. An intensive study was undertaken of the natural enemies, which were found to include several ladybird species, predatory bugs (*Orius* spp. and *Nabis* spp.) and lacewings (*Chrysopa* spp.). Three parasitic wasps (*Praon palitans* Mues., *Trioxys utilis* Mues. and *Aphelinus semiflavus* How.) were introduced. The parasites established themselves well, and in 1957 more than 100 million were collected and transferred to other fields in all the major lucerne-growing areas of California.

Sprays of parathion, mevinphos and malathion, while effective against the aphid, were found to have very deleterious effects on these natural enemies, while demeton (an organophosphorus insecticide not marketed in Britain) was less toxic. Demeton at relatively low dosage rates, while not eradicating the aphid, gave a sufficient degree of control, and as the natural enemies were little affected they were available to attack the aphids surviving the spray treatment. After ten days, for example, the number of aphids on demeton-treated plots averaged 1·5 per stem, compared with 14 on parathion-sprayed plots and 8 on those sprayed with mevinphos.

The threshold level of the aphid population at which economic crop loss begins to occur was also determined, so that by making population assessments it would be possible to decide when spraying was necessary. The natural enemies and fungal parasites were able to keep aphid infestations below the economic level for most of the year, low dosage sprays of demeton only being applied occasionally when aphid counts showed that their numbers were approaching the danger level.

The enthusiasm of many growers for integrated control methods was not apparently as high as that of the entomologists. In parts of California parathion continued to be used heavily, but with numerous failures to obtain control in 1960 and 1961; aphid resistance had increased remarkably to over a hundredfold from the fourfold level in 1956, and it was felt that the best solution to this problem would be resistant varieties of lucerne (Stern, 1962; van den Bosch, 1965).

Other pests of lucerne occur in California, such as the Egyptian alfalfa weevil, *Hypera brunneipennis* (Boh.), and any insecticides used to control them should preferably cause minimum interference with the pests' own natural enemies and those of the spotted alfalfa aphid. Tests with a range of insecticides showed that methoxychlor (an organochlorine compound) would give satisfactory control of the larvae of Egyptian alfalfa weevil and cause little mortality of most of the predators and parasites (Stern, van den Bosch and Bowen, 1962). The alfalfa caterpillars are kept in check by two native parasites supplemented, when required, by spot applications of suitable insecticides to localized outbreaks (Messenger, 1965). Screening of all new pesticides for their effects on predators and parasites has become routine in California.

An interesting complementary approach is the strip cutting of lucerne to maintain a reservoir of natural enemies in the field by avoiding the catastrophic upset occasioned by harvesting the whole crop at the same time, and to prevent the migration of a capsid bug to more susceptible crops (van den Bosch, 1965).

Another example from California of successful integrated control is that of the strawberry mite on strawberries (van den Bosch, 1965). A large number of acaricides were examined for their effects on the strawberry mite and predatory typhlodromid mites; one of them, endosulfan, was found to be highly toxic to the former but only moderately to the latter. Spectacular results are claimed following adoption of this acaricide, the strawberry mite being reduced from the status of a devastating pest to one of much less importance, with at the same time the need for fewer sprays, lower growing costs and increased yields.

An example of the possible value of strip spraying also comes from California (DeBach and Landi, 1959). In unsprayed citrus groves in Orange County, natural enemies gave satisfactory control of pests except for the purple scale, *Lepidosaphes beckii* (Newm.). A parasitic wasp was introduced which, though not giving complete control, caused sufficient mortality each year to delay population growth, so that a reduced spray programme was worth considering. After numerous trials it was found that applying an oil spray to alternate pairs of rows, leaving the intervening rows unsprayed to preserve the natural enemies, then spraying these intervening rows the next year and leaving the others unsprayed, gave a control at least as good as that usually obtained in commercial, sprayed groves and at much lower cost.

In the Netherlands, where acaricide resistance of red spider mite has become an acute glasshouse problem as it has in Britain, successful control on peaches was obtained by combining biological control with the use of selective acaricides. The biology of two predators, the ladybird *Stethorus punctillum* and the mite *Typhlodromus longipilus* Nesbitt, which commonly occur in glasshouses especially at the end of the season, and their interaction with red spider mite were studied (Bravenboer, 1959). It was considered that the rate of multiplication of these predators compared with that of red spider mite, and the amounts of prey eaten, would enable them to provide a useful if not fully economic degree of control, though this could be supplemented by the use of selective chemicals. Satisfactory control of red spider mite was obtained for three years in one peach house by taking advantage of the effects of these predators together with the use of only one spray each year of a selective acaricide, such as chlorbenside or tetradifon. In a second house, successful results were also obtained by these means for two years, then a non-selective acaricide was applied at the beginning of the third year; this killed the predators and eventually led to a rapid build-up of red spider mite which could not be controlled by six applications of various chemicals during the rest of the season.

Possibilities of integrated chemical and biological control in Britain have as yet been little explored except under glass (see page 108). The position with regard to fruit tree red spider mite has already been discussed (pages 86–87). Acaricides are sometimes required against red spider mite and strawberry mite on strawberry, and the value in this country of using chemicals which spare predatory typhlodromid mites would be worth investigating.

The view has been expressed by some workers (e.g., Turnbull and Chant, 1961) that biological control is not likely to be effective for direct pests, i.e., those damaging the part of the plant to be harvested and marketed, e.g., codling moth. The short life of certain vegetable crops and the demand for high quality produce are factors operating against biological or integrated control of vegetable pests (Dunn and Coaker, 1965). On carrots, for example, with even the high degree of parasitism of carrot fly which can occur, some damage to the roots will not be prevented. Some chemical control measure will be needed, though the insecticide should preferably have a persistence no longer than is required to cover the life of the crop. By contrast, cabbage root fly larvae feeding on the roots of cabbage and cauliflower are an indirect pest, since they are attacking part of the plant which will not be marketed, and the critical period during which protection is required seems to be the first six weeks of the plant's life. Predators can, as already described on pages 79–81, cause high mortality of the eggs and larvae of cabbage root fly, but even so this is not always sufficient to prevent loss of yield, so that some additional form of control is required. Application of insecticide as spot drenches means no more than treatment of about one hundredth of the soil surface and appears to have little effect on predator numbers. Insecticide root dips virtually leave the soil surface uncontaminated and their effectiveness is enhanced by the presence of predators. Such a method, which allows egg predation to occur, can be regarded as a step towards integrated control. Some of the less persistent organophosphorus insecticides being tested offer an advantage over dieldrin, which remains active for longer than the critical six-week period.

Aphid pests pose some particularly difficult problems. With those which are indirect pests, such as black bean aphid, a certain level of damage may be economically tolerable and advantage taken where possible of partial control by natural enemies. Where direct damage or fouling of produce is caused, as with cabbage aphid and aphids on lettuce, virtually complete control is required. In many cases aphids are also vectors of virus diseases, so that here again control by natural enemies is not likely to be adequate. With perennial crops and aphids causing indirect damage, there are probably more opportunities for making use of natural controlling factors, as has been done with the spotted alfalfa aphid in California, but similar cases are few in this country, although problems of the aphid pests of grass seed crops might be worth approaching on similar lines. Most of our aphid pests affecting fruit and vegetable crops do so for a relatively short time and, as already mentioned, there is often delay before natural enemies overtake aphids which have migrated to new crops.

In some cases breeding resistant crops may provide a solution. Even partial resistance, or breeding of plants with properties disturbing the normal behaviour pattern of a pest, could be useful in conjunction with a degree of natural control and judicious use of insecticide. Varieties of Brussels sprouts with glossy leaves have been bred which become much less infested with cabbage aphid than the normal waxy leaved varieties, though are more susceptible to the peach-potato aphid. A particularly interesting discovery in this work was that certain of the aphids' natural enemies, especially ladybirds and anthocorid bugs, preferred the glossy leaved plants, laying eggs on them but not on the waxy ones. Rove beetle and earwig predators of aphids were also occasionally found on the glossy plants but not on the waxy ones (Way and Murdie, 1965).

With certain glasshouse crops considerable attention is being given to the integration of biological and chemical control. With the demonstration of the practical value of control of red spider mite by a predatory mite (page 94) and of glasshouse whitefly by a parasite (page 98), there is a need to find chemicals which will control diseases and other pests without preventing the biological control of red spider mite and whitefly. There is also the need to produce flowers of fruit not spoiled by even small numbers of pests or their debris; it has been suggested (Hussey, 1965) that biological control methods could be used during the vegetative stages of the crop, with the use of pesticides limited to the later stages to ensure clean produce.

References and Further Reading

Bravenboer, L. (1959). De Chemische en Biologische bestrijding van de Spintmijt *Tetranychus urticae* Koch. *Publikatie van het Proefstation voor de Groenten-en Fruitteelt onder Glas te Naaldwijk*, No. 75.

CLANCY, D. W. and McALISTER, H. J. (1958). New possibilities in apple pest control. *Proc. 10th Int. Congr. Ent., Montreal, 1956*, **3**, 181–5.

DEBACH, P. and LANDI, J. (1959). Integrating chemical, biological control by strip treatment. *Calif. Citrogr.* **44**, 324, 345–7, 352.

DICKER, G. H. L. and KIRBY, A. H. M. (1955). Some impressions of apple orchard practices in Canada. *Rep. E. Malling Res. Stn for 1954*, 160–4.

DUNN, J. A. and COAKER, T. H. (1965). Rational control of vegetable crop pests. *Ann. appl. Biol.* **56**, 342–5.

EMPSON, D. W. (Ed.) (1968). Integrated pest control. Papers presented at a conference of Advisory Entomologists on 11–12 April 1967. *J. appl. Ecol.* **5**, 489–516.

F.A.O. (1966). *Proceedings of the F.A.O. symposium on integrated pest control, 1965*. Rome, F.A.O.

HUSSEY, N. W. (1965). Possibilities for integrated control of some glasshouse pests. *Ann. appl. Biol.* **56**, 347–50.

KENNEDY, J. S. (1968). The motivation of integrated control. *J. appl. Ecol.* **5**, 492–9.

MESSENGER, P. S. (1965). Utilization of native natural enemies in integrated control. *Ann. appl. Biol.* **56**, 328–30.

OATMAN, E. R. (1966). Studies on integrated control of apple pests. *J. econ. Ent.* **59**, 368–73.

PATTERSON, N. A. (1966). The influence of spray programs on the fauna of apple orchards in Nova Scotia. XVI. The long-term effect of mild pesticides on pests and their predators. *J. econ. Ent.* **59**, 1430–5.

PICKETT, A. D. and MACPHEE, A. W. (1965). Twenty years' experience with integrated control programmes in Nova Scotia apple and pear orchards. *Proc. 12th Int. Congr. Ent., London, 1964*, 597.

PUTMAN, W. L. (1963). The codling moth, *Carpocapsa pomonella* (L.) (Lepidoptera: Tortricidae): a review with special reference to Ontario. *Proc. ent. Soc. Ont. for 1962*, **93**, 22–51.

ROSS, W. A. (1953). Harmonizing chemical with biological control of orchard insects. *Rep. 13th Int. hort. Congr., London, 1952*, **1**, 293–6.

SMALLMAN, B. N. (1965). Integrated insect control. *Aust. J. Sci.* **28**, 230–4.

SMITH, R. F. and HAGEN, K. S. (1959). [The integration of chemical and biological control of the spotted alfalfa aphid.] Impact of commercial insecticide treatments. *Hilgardia* **29**, 131–54.

STERN, V. M. (1962). Increased resistance to organophosphorus insecticides in the parthenogenetic spotted alfalfa aphid, *Therioaphis maculata*, in California. *J. econ. Ent.* **55**, 900–4.

STERN, V. M., SMITH, R. F., VAN DEN BOSCH, R. and HAGEN, K. S. (1959). [The integration of chemical and biological control of the spotted alfalfa aphid.] The integrated control concept. *Hilgardia* **29**, 81–101.

STERN, V. M. and VAN DEN BOSCH, R. (1959). [The integration of chemical and biological control of the spotted alfalfa aphid.] Field experiments on the effects of insecticides. *Hilgardia* **29**, 103–30.

STERN, V. M., VAN DEN BOSCH, R. and BORN, D. (1958). New control for alfalfa aphid. *Calif. Agric.* **12** (1), 4–5, 13.

STERN, V. M., VAN DEN BOSCH, R. and BOWEN, W. R. (1962). Candidate materials to replace heptachlor as a selective control for the Egyptian alfalfa weevil in Southern California. *J. econ. Ent.* **55**, 713–18.

TURNBULL, A. L. and CHANT, D. A. (1961). The practice and theory of biological control of insects in Canada. *Can. J. Zool.* **39**, 697–753.
WAY, M. J. and MURDIE, G. (1965). An example of varietal variations in resistance of Brussels sprouts. *Ann. appl. Biol.* **56**, 326–8.
VAN DEN BOSCH, R. (1965). Practical application of the integrated control concept in California. *Proc. 12th Int. Congr. Ent., London, 1964*, 595–7.
VAN DEN BOSCH, R., SCHLINGER, E. I., DIETRICK, E. J., HAGEN, K. S. and HOLLOWAY, J. K. (1959). The colonization and establishment of imported parasites of the spotted alfalfa aphid in California. *J. econ. Ent.* **52**, 136–41.

Future Outlook

THE decade following the previous revision of this Bulletin has seen the development and widespread use of a great many new pesticides. Undoubtedly these chemicals have transformed pest control and have saved millions of pounds' worth of crops and foodstuffs which would otherwise have been lost. At the same time their use has brought problems of chemical residues in food, contamination of the environment and harmful effects on wild life, and the particularly difficult problem of the development of resistance by important pests to many of these chemicals.

While the attitude of some economic entomologists has been, as Jacob (1958) put it, that 'salvation lies at the end of a spray nozzle' and that the chemists will always be ready to produce a new pill to correct shortcomings or failures of previous ones, all biologists do not share such views. Not only is the production of new pesticides very costly but it is being increasingly realized that the mere application of toxic chemicals does not necessarily provide a satisfactory, and in the long term successful, method of pest control. We have been gradually forced to recognize the adaptability of living organisms and their ability as an evolutionary process to meet new pressures in their environment. We can see from Smallman's definition of 'integrated control' on page 104 that we are faced with an evolutionary race between the responsive powers of the pest to threats to its survival on the one hand and on the other our ability to develop control techniques. In this race we must handicap the pest by devising a variety of techniques so that relatively mild selection pressures can be applied more than one at a time.

Some biologists argue for greater diversity of cropping and hence of the associated animal life. In a diverse fauna natural regulation tends to a balance between populations of the various animals, with no one species achieving a consistent or lasting advantage. But this relative stability is only long term and, as pointed out in the Introduction, natural regulation does not prevent devastating population 'explosions' of one insect or another from time to time. In any case the growing of crops creates a balance in favour of one plant species and inevitably of insects living on it, and the agriculturalist insists that there is no going back: for economic reasons modern agriculture must develop towards more and more large units specializing in certain crops, and the ecology of these units will change with advances in husbandry techniques. There will still be a place for pesticides but used critically, at the right place and the right time, and in the context of carefully worked out programmes which include a variety of other techniques to tip the balance against the pest. Theobald (1909), writing of the 'endless loss' to fruit growers caused by pests, said 'Man must destroy his enemies, and it is useless to rely on Nature to do so for him', but he would probably agree that more subtle methods than the use of the spraying machine alone are imperative now.

The natural enemies of insects and mites are of great potential importance and, though advantage may not always need to be taken of them in systems of pest control, the provision of suitable habitats for the maintenance of useful species needs to be considered. Modern agriculture is decreasing the diversity of the flora and fauna of the countryside and, although conservation of wild life in general is outside the scope of this Bulletin, thought also needs to be given to the requirements of those other beneficial insects, the pollinators of blossom.

References and Further Reading

JACOB, F. H. (1958). Some modern problems in pest control. *Sci. Prog., Lond.* No. 181, 30–45.

KENNEDY, J. S. (1968). The motivation of integrated control. *J. appl. Ecol.* **5**, 492–9.

THEOBALD, F. V. (1909). *The insect and other allied pests of orchard, bush and hothouse fruits and their prevention and treatment.* Wye, F. V. Theobald.

Printed in England for Her Majesty's Stationery Office, by Elliott Bros. & Yeoman Ltd., Liverpool
3278 Dd. 147079 5250 3/70 Gp 474

Index

Index

Index

Printed and published by
HER MAJESTY'S STATIONERY OFFICE

To be purchased from
49 High Holborn, London W.C.1
13A Castle Street, Edinburgh EH2 3AR
109 St. Mary Street, Cardiff CF1 1JW
Brazennose Street, Manchester M60 8AS
50 Fairfax Street, Bristol BS1 3DE
258 Broad Street, Birmingham 1
7 Linenhall Street, Belfast BT2 8AY
or through any bookseller

Printed in England

Dd. 147079 7/69

9780112403203